Cambridge International AS & A Level

Physics
Practical Skills

Catherine Jones

Cambridge International copyright material in this publication is reproduced under licence and remains the intellectual property of Cambridge Assessment International Education.

The practice questions, accompanying marks and mark schemes included in this resource have been written by the author and are an opportunity to practise exam skills and are a guideline only. They do not replicate examination papers. In examinations, the way marks are awarded may be different. Any references to assessment and/or assessment preparation are the publisher's interpretation of the syllabus requirements and may not fully reflect the approach of Cambridge Assessment International Education.

Third party websites and resources referred to in this publication have not been endorsed by Cambridge Assessment International Education.

The Publishers would like to thank the following for permission to reproduce copyright material.

Photo credits

p. 11 © Epnov/stock.adobe.com; p. 16*tl* © Note_yn/stock.adobe.com; p. 16*tr* © Andrei Nekrassov/stock.adobe.com; p. 16*b* © Pioneer111/stock.adobe.com; p. 62 © Michaklootwijk/stock.adobe.com

Every effort has been made to trace all copyright holders, but if any have been inadvertently overlooked, the Publishers will be pleased to make the necessary arrangements at the first opportunity.

Although every effort has been made to ensure that website addresses are correct at time of going to press, Hodder Education cannot be held responsible for the content of any website mentioned in this book. It is sometimes possible to find a relocated web page by typing in the address of the home page for a website in the URL window of your browser.

Hachette UK's policy is to use papers that are natural, renewable and recyclable products and made from wood grown in well-managed forests and other controlled sources. The logging and manufacturing processes are expected to conform to the environmental regulations of the country of origin.

Orders: please contact Hachette UK Distribution, Hely Hutchinson Centre, Milton Road, Didcot, Oxfordshire, OX11 7HH. Email education@hachette.co.uk. Telephone: +44 (0)1235 827827. Lines are open from 9 a.m. to 5 p.m., Monday to Friday. You can also order through our website: www.hoddereducation.com

ISBN: 978 1 5104 8284 5

© Catherine Jones 2023

First published in 2023 by
Hodder Education,
An Hachette UK Company
Carmelite House
50 Victoria Embankment
London EC4Y 0DZ

www.hoddereducation.co.uk

The authorised representative in the EEA is Hachette Ireland, 8 Castlecourt Centre, Dublin 15, D15 XTP3, Ireland (email: info@hbgi.ie)

Impression number   10 9 8 7 6 5 4 3 2 1

Year   2027 2026 2025 2024 2023

All rights reserved. Apart from any use permitted under UK copyright law, no part of this publication may be reproduced or transmitted in any form or by any means, electronic or mechanical, including photocopying and recording, or held within any information storage and retrieval system, without permission in writing from the publisher or under licence from the Copyright Licensing Agency Limited. Further details of such licences (for reprographic reproduction) may be obtained from the Copyright Licensing Agency Limited, www.cla.co.uk

Typeset in India

Printed in the UK

A catalogue record for this title is available from the British Library.

Contents

Introduction	4
1 Safety	5

AS Level Practical Skills — 7

2 Manipulation, measurement and observation — 7
  2.1 Successful collection of data: Following instructions — 7
  2.2 Successful collection of data: Errors — 8
  2.3 Successful collection of data: Uncertainty — 10
  2.4 Successful collection of data: Choosing instruments — 11
  2.5 Quality of data — 17

3 Presentation of data and observations — 19
  3.1 Tables — 19
  3.2 Graphs — 21

4 Analysis, conclusions and evaluation — 25
  4.1 Interpreting graphs — 25
  4.2 Estimating uncertainties — 29
  4.3 Combining uncertainties — 30
  4.4 Drawing conclusions — 32
  4.5 Identifying limitations and suggesting improvements — 35

5 Practice questions — 38

A Level Practical Skills — 58

6 Planning — 58
  6.1 Defining the problem — 58
  6.2 Methods of data collection — 59
  6.3 Methods of analysis — 64
  6.4 Considering safety — 66

7 Analysis, conclusions and evaluation — 68
  7.1 Data analysis — 68
  7.2 Tables of results — 70
  7.3 Graphs — 71
  7.4 Treatment of uncertainties — 72
  7.5 Conclusion — 78

8 Practice questions — 80

Introduction

This workbook aims to provide you with the practical skills and knowledge required for the Cambridge International AS & A Level Physics syllabus (9702). Collecting accurate data is essential to the development of physics ideas through a process known as the scientific method. It is the key to knowing if a theory is valid. This workbook is designed to enable you to use a wide variety of physics equipment, design investigations that yield accurate results, analyse data, draw conclusions and understand the quality of your evidence.

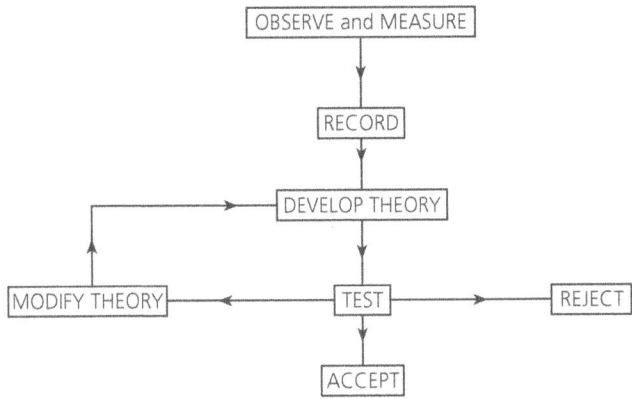

Figure 0.1 Block diagram to illustrate the scientific method

The first chapter looks at safe working procedures in the laboratory. The book is then divided into two sections. Chapters 2 to 5 cover the practical skills required at AS Level and Chapters 6 to 8 cover the additional skills required for the full A Level qualification. The initial chapters focus on the skills specified by the practical assessment section of the syllabus. For AS Level these are 'Manipulation, measurement and observation', 'Presentation of data and observations' and 'Analysis, conclusions and evaluation'. For the full A Level these are 'Planning' and 'Analysis, conclusions and evaluation' (this further develops the skills from AS Level). These chapters contain guidance, examples and exercises for you to practise the skills outlined. There are spaces for you to write your answers in the book.

Chapters 5 and 8 feature practice questions taken from previous Cambridge International AS & A Level Physics syllabus (9702) examinations. Information about which examination series they are sourced from is given at the end of the question. There are spaces for you to write your answers and include calculations. Answers are written by the author and are available at **www.hoddereducation.com/cambridgeextras**.

Assessment overview

There are two components of the assessment designed to test your experimental skills:
- Paper 3 Advanced Practical Skills: In this laboratory-based paper you will be required to carry out two investigations and answer questions on them. The skills required for this paper are covered in Chapters 2 to 4. All students take this paper.
- Paper 5 Planning, Analysis and Evaluation: In this written paper you will be required to plan an investigation and to analyse the results of a second investigation. The skills required for this paper are covered in Chapters 6 and 7. Only students studying the full A Level qualification take this paper.

AS & A LEVEL

1 Safety

In any practical it is very important to work safely. Your teachers will have carried out a risk assessment for any practical work that you do. However, it is your responsibility to read the instructions and listen to any verbal warnings they may give regarding the safety of a procedure.

For the A Level qualification, you will be expected to identify risks and suggest any precautions you must take to minimise those risks. However, it is always good practice to take responsibility for your own safety.

When conducting any experiment, start by reading through the method and imagine setting up the apparatus. Then consider the possible risks of that procedure. For example, consider an investigation into the extension of a metal wire under increasing masses. The apparatus is shown in Figure 1.1.

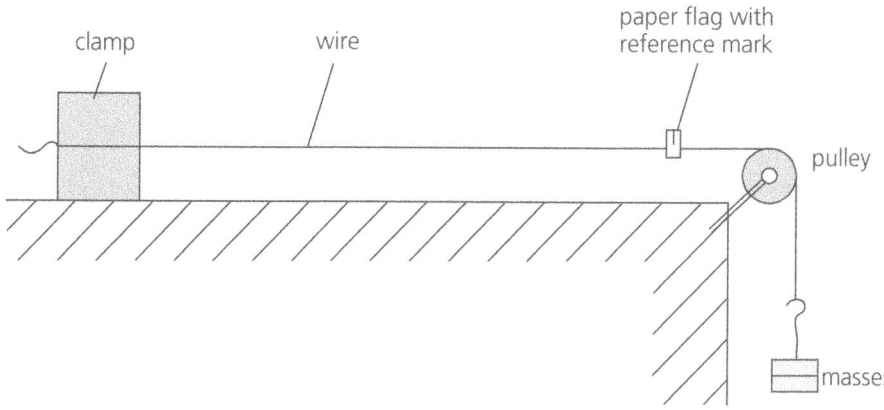

Figure 1.1 Simple experiment to measure the Young modulus of a wire

1 Identify any hazards for the experiment.

   A hazard is something that may cause you harm. In this investigation the metal wire and the masses used are a possible hazard.

2 Consider the risk of that hazard.

   The risk is the chance that the hazard will cause you harm. You need to think what the possible harm is, how likely it is and how serious the effects could be. The risk from the metal wire is that it could break suddenly, move quickly and hit you. How likely this is will depend upon the masses used. If this did happen and it hit your eye the harm would be significant.

3 Identify suitable precautions.

   Precautions are the steps you take to reduce the risk to yourself. Wearing safety glasses protects your eyes; taking measurements from a suitable distance and observing the wire carefully reduce the chance of harm to your face.

AS & A LEVEL

EXERCISE 1A

In the investigation into the extension of a metal wire under increasing masses, identify the risk from the masses and suggest suitable precautions you could take.

Risk: ....

....

Precaution: ....

....

Some general rules for safety in a physics laboratory include:
- know where the first aid kit and eye wash kit are located in the classroom
- make sure you know what to do in case of a fire and the locations of fire safety equipment such as fire extinguishers and/or fire blankets
- read your instructions carefully and ask questions if you are not sure about anything
- read, understand and follow all warning labels for specific hazards:

Figure 1.2 Hazard symbols. **a)** Corrosive; **b)** Harmful to the environment; **c)** Flammable; **d)** Oxidising; **e)** Toxic; **f)** Irritant; **g)** Radioactive; **h)** Laser

- never eat, drink or smoke in a laboratory
- wear sensible clothing and tie back long hair
- always set up apparatus away from the edge of the bench and work in a logical and tidy manner
- always wear eye protection when working with chemicals or heating water in glassware
- do not allow electrical equipment to come into contact with water
- report any accidents to your teacher and clean up quickly as per instructions
- at the end of any practical work, leave your working area clean and tidy and wash your hands.

AS LEVEL 2

Manipulation, measurement and observation

In this chapter, you will look at the process of collecting data. It is important you understand how to take accurate measurements and choose the most suitable measuring instruments.

2.1 Successful collection of data: Following instructions

For any practical work, you will be given a series of instructions for you to follow. It is good practice to:
- read through *all* the instructions carefully to make sure you know what is expected of you
- consider any safety aspects that arise from the practical
- prepare tables to record your results, including columns for repeats and any analysis you may want to do
- note down any difficulties you have with making measurements as these may explain any anomalies you have.

You will be expected to understand diagrams of apparatus and to follow circuit diagrams to build electrical circuits.

Building circuits from circuit diagrams

It is important that you know and can recognise all the different circuit symbols.

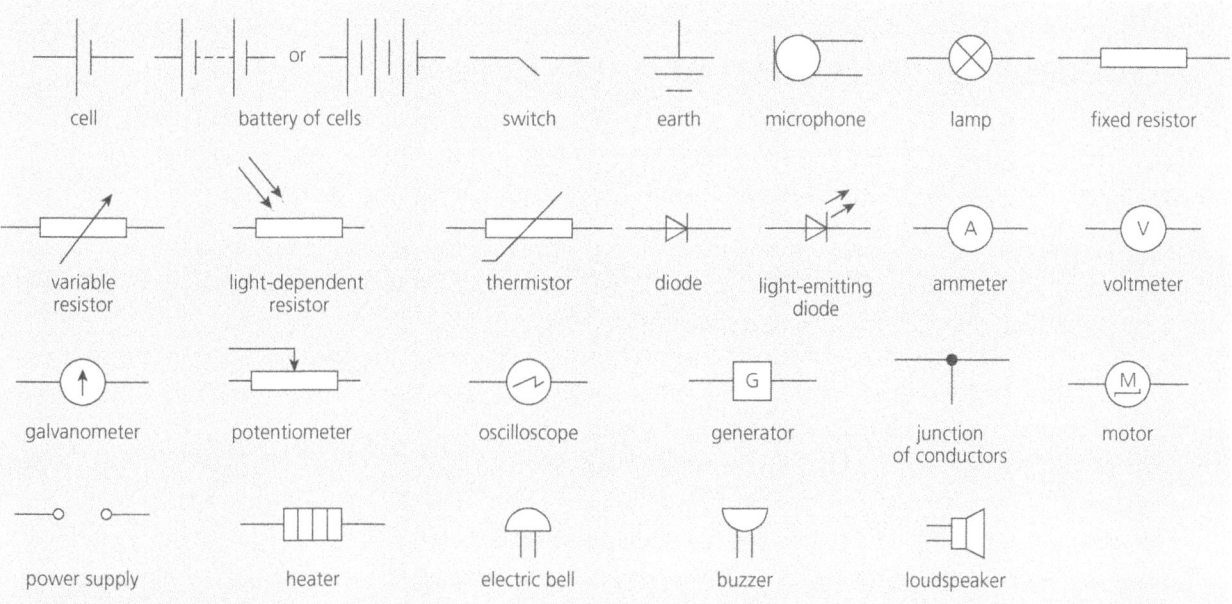

Figure 2.1 Circuit symbols

When building a circuit, start at the power source. Starting with the positive terminal of the power source, follow a single loop and add the components in turn until you are back at the negative terminal of the supply. If there is more than one loop add that after the first is completed. Always add voltmeters at the end as these are placed in parallel with the part of the circuit or component they are measuring the voltage across.

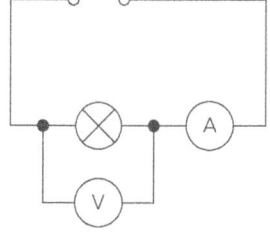

Figure 2.2 A simple circuit diagram

For example, to build the circuit in Figure 2.2 you need a lead to join the positive terminal of the power supply to the lamp, a lead to join the lamp to the positive terminal of the ammeter and a lead to join the negative terminal of the ammeter to the negative terminal of the power supply. You can then use two leads to place the voltmeter across the lamp, as shown in Figure 2.3.

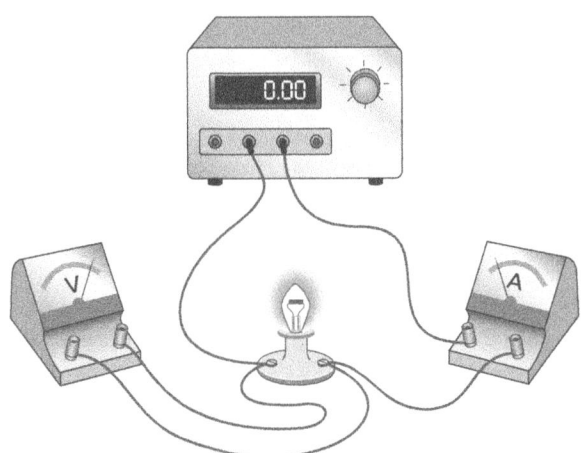

Figure 2.3 A completed circuit using standard equipment

Circuits are often more complex than this, but if you follow the diagram for each loop, you will be able to construct them.

2.2 Successful collection of data: Errors

In any measurement you are attempting to read the **true** value. An accurate measurement is one which is close to the true value. It is important to remember that with any instrument you use you will get measurement error. This is the difference between the true value and the measured value.

There are errors in any experiment. These can be errors in the instrument itself, in reading the instrument or in the design of the experiment. There are two main sources of error for you to be aware of whenever you take a measurement: these are *random* and *systematic* errors.

1. Random errors

    Random errors can happen in any measurement and, as the name suggests, they vary unpredictably. Some examples of random errors are:
    - reading a scale at different angles and introducing a parallax error; this is where the reading is higher or lower than the true value (see Figure 2.4)
    - difficulties reading an analogue scale when the value lies between the marks on a scale (see Figure 2.5)
    - timing an event and knowing exactly when to start and stop a stopwatch and errors introduced by your reaction time
    - difficulties in taking two measurements at the same time such as the temperature at set time intervals.

    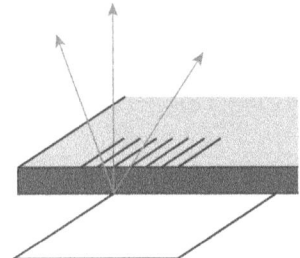

    Figure 2.4 Parallax errors when reading a ruler

    Figure 2.5 Difficulties reading an analogue scale

    You cannot correct random errors but you can reduce their effect by repeating and finding average values and by plotting a graph and drawing a line of best fit. Random errors affect the precision of your results. Precise results are ones where there is little spread around the mean value and points lie close to the line of best fit.

2 Systematic errors

Systematic errors are such that the measured value is shifted the same amount from the true value every time the measurement is taken. Repeating the reading does not reduce this error. Some examples of systematic error are:
- zero error; this is when the measuring instrument is not at zero before taking the measurement, for example, forgetting to zero or tare a top pan balance before finding the mass of something.
- the instrument may be calibrated incorrectly, for example, a newton meter may have been stretched beyond its elastic limit.

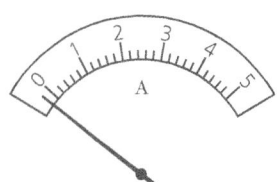

Figure 2.6 This ammeter has a zero error of about −0.2A

If you know the value of the systematic error it may be possible to correct for it. You can do this by either adding or subtracting it from your measured value.

EXAMPLE 2.1

A micrometer reads 0.06 mm when it should read zero. When measuring an object the reading is 2.53 mm. Determine the corrected reading.

Method

The corrected reading is:

Corrected reading = reading − zero error.

Corrected reading = 2.53 − 0.06 = 2.47 mm

EXERCISE 2A

For each example, decide whether it is most likely to involve random or systematic error. Explain your answers.

a In an experiment to investigate how the length of a pendulum affects the period, a student measured to the top of the pendulum bob instead of the middle. [2]

....

....

b A student is monitoring the temperature of a liquid as it is heated. They read the thermometer every 30 seconds. [2]

....

....

c Measuring the time it takes a parachute to fall 1.50 m. [2]

....

....

AS LEVEL

d Measuring the extension of a spring as you increase the load. [2]

....

....

2.3 Successful collection of data: Uncertainty

The uncertainty in any measurement is the range of values between which the true value can be expected to lie. There are a number of ways of determining the uncertainty in any measurement and you will cover this in Chapter 4. In this chapter, you will consider the uncertainty in the measuring instrument itself.

Some instruments have their uncertainty marked on them. For example, digital meters have an uncertainty quoted as 1.5% ± 2 digits. The 1.5% refers to the reading on the scale and ± 2 digits is the uncertainty in the final figure of the reading. For example, a reading of 2.45 A would have an uncertainty of

$$\left(\frac{1.5}{100} \times 2.45\right) \pm 0.02 = \pm 0.06$$

If this is not the case or if you do not have the information, a simple way of identifying the uncertainty is to look at the smallest change in the quantity the instrument can measure. For example, a metre ruler has an uncertainty of ± 1 mm and a top pan balance that reads mass to two decimal places has an uncertainty of ± 0.01 g.

For some measurements you can reduce uncertainty by measuring multiple instances. The uncertainty in the measurement then becomes the uncertainty divided by the number of instances. For example, to find the mass of one ball bearing you can measure the mass of 10 and divide by 10. If the mass of 10 is 0.33 g ± 0.01 g, then one ball bearing has a mass of 0.033 ± 0.001 g.

You can express uncertainty as an absolute uncertainty (0.033 ± 0.001 g or as a percentage (0.033 ± 3%).

> **TIP**
> To calculate the percentage uncertainty in reading x if the uncertainty is Δx:
> $$\text{Percentage uncertainty} = \frac{\Delta x}{x} \times 100$$

When choosing the instrument think about the uncertainty of the instrument and your requirements. For example, if you wish to measure the thickness of a page in this book a 30 cm ruler is not appropriate. The smallest interval is 1 mm and this page is probably between 0.05 and 0.10 mm. A 30 cm ruler would also not be the right choice for measuring 2 mm. This is because the uncertainty in the measurement would be 50% of the reading. In each case, you would need to choose an instrument with a smaller uncertainty such as a micrometer or calipers.

EXERCISE 2B

a State the uncertainty in each of the following instruments from the readings shown. [4]

  i 12.4 °C

....

  ii 12.50 g

....

iii 0.26 mm

iv 45°

b A voltmeter is quoted as having an uncertainty of 2% ± 3 digits. The reading is 2.58 V. Calculate the uncertainty in the reading. [1]

2.4 Successful collection of data: Choosing instruments

This section looks at how you choose the appropriate instrument for each measurement and how to use it.

2.4.1 Measuring length

1 Choosing the correct instrument

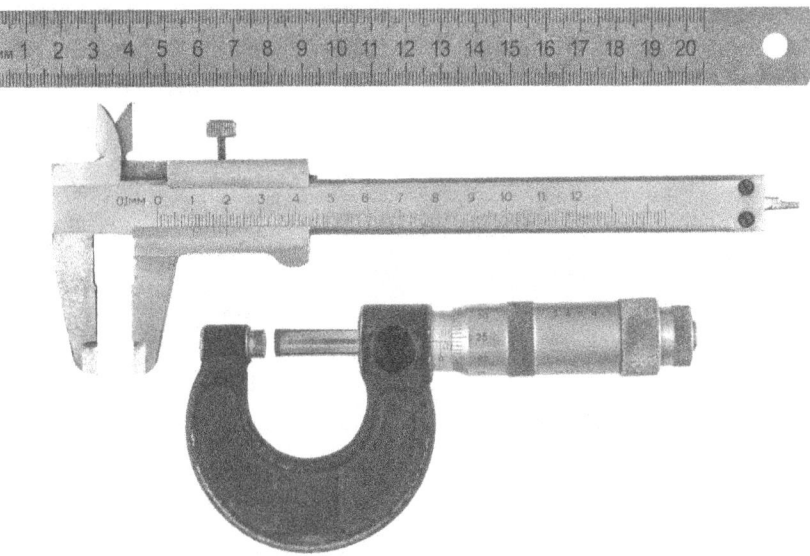

Figure 2.7 Ruler, calipers and micrometer

Figure 2.7 shows different measuring instruments each with a different uncertainty. When choosing the instrument consider the uncertainty you need for your measurement and the upper limit of that instrument.

- Micrometer screw gauge: uncertainty ± 0.01 mm and upper limit 25 mm
- Calipers: uncertainty ± 0.1 mm and upper limit 20–30 cm
- Metre ruler: uncertainty ± 1 mm and upper limit 1 m
- Tape measure: uncertainty ± 1 mm and upper limit 10 m.

The measuring instrument should always have an upper limit greater than the reading. For example, to measure a length greater than one metre you should use a tape measure rather than two metre rulers. To measure the diameter of a steel ball or a wire or the thickness of paper you will probably use calipers or a micrometer screw gauge.

AS LEVEL

2 How to use

For the micrometer, place the object you wish to measure in the gap and tighten using the ratchet (dial) at the end. When you hear a click stop tightening.

For the calipers, place the object between the external or internal jaws. The internal jaws are useful for measuring the diameter inside a test tube for example. Close slowly using the dial. Stop as soon as you feel any resistance.

When using a micrometer screw gauge or vernier calipers always make sure they are zero when they are closed. It is possible to zero both of these instruments. However, if that is not possible, then record the zero error and use it to correct your measurements.

3 How to read

Micrometers always have vernier scales and calipers can have vernier or digital scales. A vernier scale consists of a main scale and a moving second scale. They are used to make very precise measurements.

Figure 2.8 shows a linear vernier scale on a pair of vernier calipers.

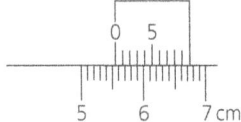

Figure 2.8 Linear vernier scale

To read a linear vernier scale start by reading the main scale. Look at the last whole mark visible before the 0 mark. Then look at the second scale and find the mark that lines up best with the main scale. Add the two values together to get the reading.

Figure 2.9 shows a circular vernier scale on a micrometer screw gauge.

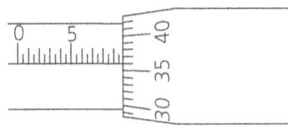

Figure 2.9 Circular vernier scale

To read a circular vernier scale, start with the main scale. Read off the last whole or half mark visible. Then read from the second scale. This reading will be the value of the second scale that lines up with the centre of the main scale. Add the two values together to get the reading.

EXAMPLE 2.2

Explain how to take the reading on

a the vernier calipers in Figure 2.8

b the micrometer screw gauge shown in Figure 2.9.

Method

a In Figure 2.8 the main scale has millimetre intervals. The last whole mark visible before the 0 mark is 5.5 cm. The second scale gives the reading to the nearest 0.1 mm. The second scale mark that best lines up with the main scale is the third or 0.03 mm. Therefore, the reading is 5.53 cm or 55.3 mm.

b In Figure 2.9, the main scale is marked with whole and half millimetres. The last mark visible is 9.5 mm. The second scale is marked in 0.01 mm intervals. The value on the second scale that lines up with the centre of the main scale is 36 or 0.36 mm. Therefore, the reading is 9.86 mm.

EXERCISE 2C

Figure 2.10 shows two vernier scales. Determine the reading in each case. [2]

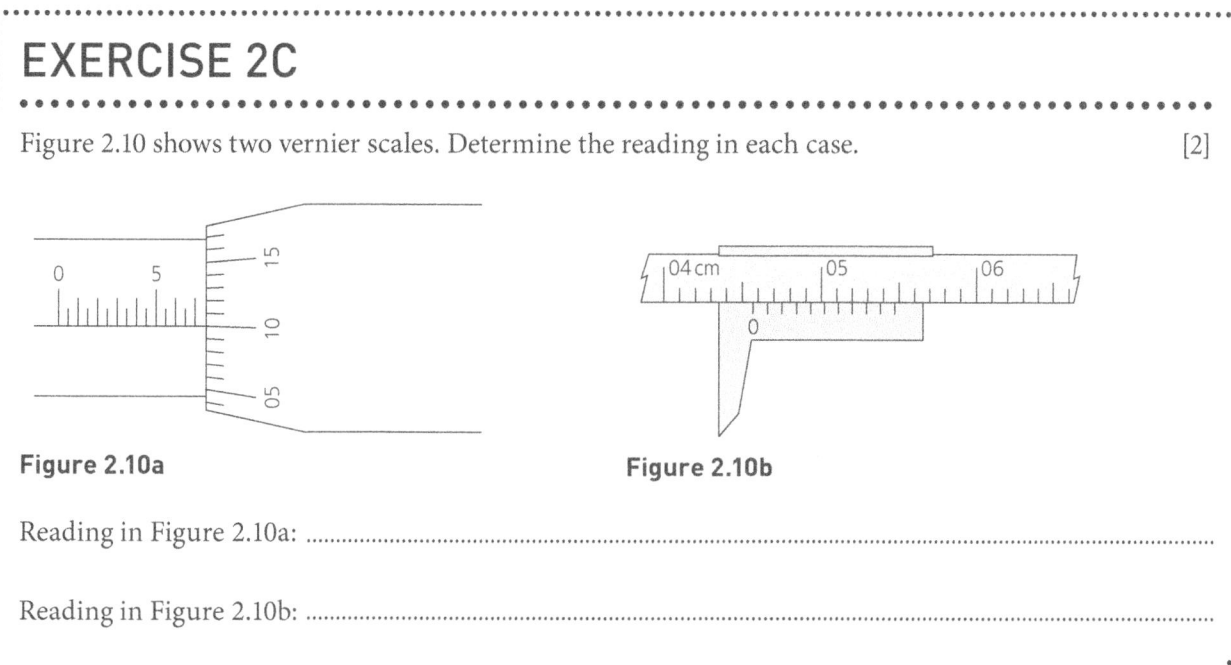

Figure 2.10a **Figure 2.10b**

Reading in Figure 2.10a: ....

Reading in Figure 2.10b: ....

2.4.2 Measuring time

The smallest interval measured by many digital stopwatches is 0.01 s. However, when you time an event there will always be an error due to your reaction time. Typically, human reaction time is 0.2 s. With practice at timing an event you may be able to reduce this, but at best it will be 0.1 s. This gives an uncertainty ± 0.1 s to your measurements. To reduce the impact of your reaction time always try to measure longer time intervals so that the uncertainty is less important. As reaction time is a random error, you should always repeat measurements.

Figure 2.11 shows two oscillating systems; a pendulum and a mass on a spring.

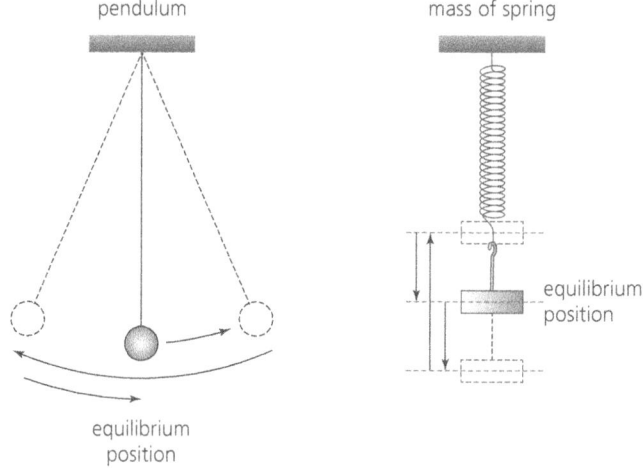

Figure 2.11 Two oscillating systems

To time one complete oscillation, time 10 complete oscillations and divide by 10. Repeat this process. Place a pin or marker in the equilibrium position so that you have a reference point from which to start and stop your timing.

In motion experiments, light gates can be used to measure time. Light gates have an infrared transmitter and receiver. As the object passes through the light gate, it blocks the infrared beam and starts the counter. The counter stops when the beam is detected again at the receiver.

AS LEVEL

2.4.3 Measuring temperature

Temperature is measured using a thermometer, which can either be liquid-in-glass or digital. It is important that the thermometer is in good thermal contact with the object being measured. Wait until the thermometer and object are in thermal equilibrium (the temperature will be constant) before you take your reading.

Most liquid-in-glass thermometers have intervals equal to one degree Celsius and so have an uncertainty of ± 1 °C. Digital thermometers often have an uncertainty of ± 0.1 °C. One advantage of a digital thermometer is that you can connect it to a data logger and record the temperature as the data is collected. This is a good choice of measuring instrument when the temperature is changing rapidly with time, for example, if you were monitoring the temperature change during a change of state.

When reading a liquid-in-glass thermometer make sure your eye is level with the thermometer and the thermometer is vertical when you take your reading. This is shown in Figure 2.12.

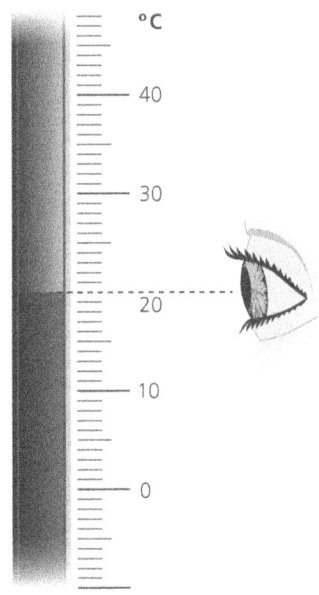

Figure 2.12 How to read a thermometer

2.4.4 Measuring mass

Top pan balances each have a maximum mass that they can measure. It is important that you do not overload the balance. For example, a balance that reads to two decimal places often has an upper limit of 200 g and an uncertainty of ± 0.01 g.

If you simply want to find the mass of an object for your experiment:
- make sure the pan is clean
- tare (zero) the balance
- carefully place the object on the pan. If it is a large object, then it must be in contact with only the pan.

If you wish to measure the mass of a powder or liquid, place a container on the scales and tare the balance and then add the substance to the container.

2.4.5 Measuring angles

To measure angles, you use a protractor. Figure 2.13 shows the important features of a protractor. This protractor has an uncertainty ± 1°.

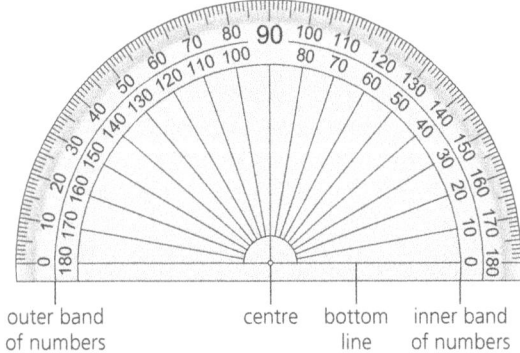

Figure 2.13 The key features of a protractor

2 Manipulation, measurement and observation

To measure angle AOX in Figure 2.14:
- Align the bottom line on the protractor with the line AB and the centre of the cross with the point O.
- Use the outer band for this angle as it begins with zero.
- You can clearly see that the angle is 55°.

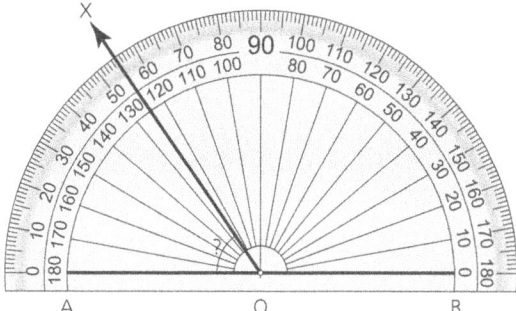

Figure 2.14 Measuring angle AOX

To measure the angle XOB, you would use the inner band. In Figure 2.14, the lines are drawn thickly so that you can easily see them. When you draw lines, they should be drawn using a ruler and a *sharp* pencil. This reduces the error in measuring the angle as thick lines increase the uncertainty in your measurement.

2.4.6 Measuring force

A newton meter can be used to measure force. Newton meters contain a spring which has been calibrated so that the force needed for a particular extension is recorded on the scale. Always zero the newton meter before you take a reading.

When you weigh an object the newton meter must be hung vertically. Use the hook to suspend it from a clamp stand. Figure 2.15 shows a newton meter being used to measure the weight of an apple.

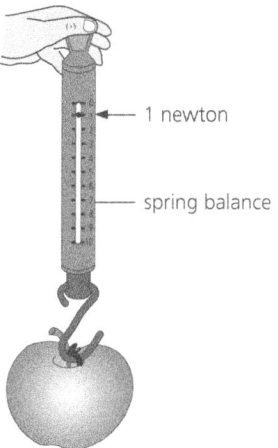

Figure 2.15 A newton meter measuring the weight of an apple

Newton meters have different ranges. If you are unsure of the force, start with a newton meter with a high upper limit and then move down until you have one with the appropriate upper limit and least uncertainty. For example, if it had been available, a 0–5 N newton meter would have been a better choice to measure the apple in Figure 2.15.

2.4.7 Measuring potential difference and current

Potential difference is measured using a voltmeter connected in parallel with the component. Current is measured using an ammeter placed in series with the component. Voltmeters and ammeters can be:
- analogue where a moving needle indicates the value
- digital where the value is shown as digits on a screen.

Conventional current is from positive to negative so always connect your meter so that the positive end of the meter is connected to the positive side of the power source.

Regardless of which electrical meter you use, if you do not know the expected voltage or current reading it is always best practice to start on the largest scale then move down through them.

Figure 2.16 Two different analogue ammeters

For example, to measure an unknown current you would start with the ammeter on the left. A full-scale deflection of the needle on this ammeter is 10.0 A and the uncertainty is ± 0.2 A. If the current reading was less than 2.0 A use the ammeter on the right instead. This has a full-scale deflection of 2.00 A and the uncertainty is ± 0.05.

You may have access to a digital voltmeter or ammeter or to a multimeter. Figure 2.17 shows a multimeter.

Figure 2.17 Multimeter set to read up to 200 V maximum; the uncertainty in reading will be ± 0.1 V

To use the multimeter, always connect one lead to the input labelled 'COM'.

If you wish to use it as a *voltmeter* connect the second lead to the input labelled 'V Ω mA'. You then set the dial to the appropriate scale either DC voltage (choice of 200 mV, 2 V, 20 V, 200 V or 1000 V) or AC voltage (choice of 2 V, 20 V, 200 V or 750 V) on the multimeter in Figure 2.17. The lower the maximum voltage, the smaller the interval it can measure. If the value is higher, the meter will simply show 1. Choose the scale with the smallest interval possible for your reading.

If you wish to use as an ammeter, then always start with the second input in '20 ADC' and turn the dial to 20 A (smallest interval is 0.1 A). If the current is below 0.2 A, then you can change the input to the 'mA' and choose a different current setting. For example, with a setting of 20 mA, the maximum current is 20 mA (0.02 A). On this scale, the smallest interval is 0.01 mA.

EXERCISE 2D

Density is defined as the mass per unit volume. Describe the measurements you would take to determine the density of a wooden cube. Include the uncertainty in each measurement. [6]

....

....

....

....

....

....

....

2.5 Quality of data

To improve the quality of the data you collect, you have to consider carefully the number of repeats, and the number and the range of measurements you will take.

2.5.1 Repeats

It is always good practice to repeat measurements. Repeating measurements reduces the uncertainty caused by random errors. For example, when measuring the diameter of a wire, measure in a number of places along the length and then calculate the mean. This is because the wire may not be of uniform thickness and the diameter may vary along the length.

2.5.2 Number of measurements

To decide the number of measurements, you need to consider how you will process the data. For example, if you will be plotting a graph and you expect a straight-line trend, then six sets of measurements will be sufficient. However, if the graph is curved you may need to take more measurements to determine the relationship.

> **TIP**
> In the AS practical assessment, the results are often used to plot a straight-line graph. So, in these assessments, always collect six sets of measurements.

2.5.3 Range

When deciding on the range for your experiment, you need to take measurements over the largest range possible for your available equipment or for the instructions you have been given. For example, you might be asked to investigate the behaviour of a spring under increasing load. You have been given ten 100 g masses. You could therefore have a range from 100 to 1000 g. However, the instructions say you are not to use more than 600 g or you will damage the spring. Therefore, the maximum possible range you can investigate is 100 to 600 g, increasing in 100 g increments.

AS LEVEL

It is usual to increase the independent variable in equal steps over the whole range. There may be occasions in an investigation where when you look at the graph, you want to investigate behaviour in a certain region. In that case, simply take further measurements in the place you are interested in.

> **TIP**
> Only put the apparatus away after you have drawn your graph or analysed your data. This allows you to investigate whether any anomalies or turning points resulted from the way you set up your apparatus.

EXERCISE 2E

A student is investigating how the temperature of a liquid determines the time it takes a ball bearing to fall through the liquid.

a State the measurements the student would need to take and the most suitable instruments to use. [4]

....

....

....

....

b Identify sources of possible error in the experiment. [3]

....

....

....

....

c Suggest a suitable number of measurements, range and number of repeats. [4]

....

....

....

....

AS LEVEL 3

Presentation of data and observations

In this chapter, you will look at how to record and present your results. It is important that you present your measurements and observations clearly. There is an agreed format all scientists use which is explored in this section.

3.1 Tables

3.1.1 Designing tables

Tables are used to display results. Tables should be drawn with a pencil and ruler before you begin collecting results. There should be columns to include all the raw data, a column for each repeat and a column for the average. Include columns for any calculations you need to do.

Each column should include a heading that will tell you the quantity in the column and its unit. The quantity and the unit can both be written as words or symbols. The convention is to write the heading as quantity/unit, for example, Force/newton or F/N.

EXAMPLE 3.1

You have been asked to investigate how the depth, d, of water affects the speed, v, of a wave. You have been given a tray, a ruler and a stopwatch. You will time a wave as it travels from one end of the tray and back. Design a suitable table to record your results.

Method

Distance travelled by the wave = _____ m

d/mm	t/s			v/m s^{-1}
	1	2	Average	

EXERCISE 3A

A student is investigating how the resistance of a component changes with increasing potential difference. They use a voltmeter and ammeter to make the measurements. Draw a suitable table for them to record their results in, including repeats. [2]

AS LEVEL

3.1.2 Recording measurements

All raw data in a column should be written to the same number of decimal places. This is because the number of decimal places tells you the uncertainty in the reading. For example, if you are using a ruler, you can measure each result to the nearest millimetre. Therefore, all the results are recorded to the nearest millimetre, for example, 25.4 cm, 30.0 cm, 40.0 cm. If you recorded the last result as simply 40 cm it would imply that you only knew the result to the nearest cm.

3.1.3 Calculated values in tables

When you record calculated values in tables the number of significant figures depends on the value used to calculate it with the fewest number of significant figures. The number of significant figures will be the same as (or one more than) the number with the fewest number of significant figures used to calculate it.

For example, you are asked to calculate the power, P, and know that the current, I, is 0.45 A and the potential difference, V, is 3.45 V.

$P = VI = 1.5525$ W

In this case, you know the potential difference to 3 significant figures and the current to 2 significant figures. Therefore, you can quote the power to 2 significant figures (or 1 more).

$P = 1.6$ W or $P = 1.55$ W

Sometimes you have to determine the unit of your calculated value. Perhaps you have been asked to determine the velocity squared, v^2, or the inverse of the displacement, $\frac{1}{s}$. The column heading for v^2 would be $v^2/\text{m}^2\text{s}^{-2}$. The column heading for $\frac{1}{s}$ would be $\frac{1}{s}/\text{m}^{-1}$.

EXAMPLE 3.2

In the speed of the wave investigation from Example 3.1 you are told that

$v = \sqrt{gd}$

To investigate this relationship, you include a column in your table for the square root of the depth $\sqrt{d}$.

The heading and the units for this column would be: $\sqrt{d}/\text{mm}^{\frac{1}{2}}$ or $d^{\frac{1}{2}}/\text{mm}^{\frac{1}{2}}$.

EXERCISE 3B

a A student is investigating how the diameter of a wire affects its resistance. They measure the diameter, d, using a micrometer and use a voltmeter and ammeter to determine the resistance. They have been asked to determine $\frac{1}{d^2}$. They are not planning to repeat. Draw a suitable table for them to record their results in, including calculated values. [2]

b A student takes measurements to determine the spring constant, k, of a material. They know the mass to the nearest gram and the extension to the nearest mm. Their results are as follows:

Mass, m/g	Force, F/N	x/cm	k
100	0.981	1.7	57.71
200	1.962	3.6	54.5
300	2.943	6	49.05
400	3.924	7.7	50.96

Complete the table and correct any errors $\left(k = \frac{F}{x}\right)$. [3]

TIP
Raw data is often written to the same number of decimal places. The measuring instrument decides this. Calculated data is usually written to the same number of significant figures. However, it may vary as the number is decided by the significant figures of the values it is calculated from.

3.2 Graphs
3.2.1 Drawing the axes

In an experiment, the independent variable (the one you change in the experiment) is generally on the x-axis and the dependent variable (the one that changes as a result) on the y-axis.

When drawing a graph, label both the x-axis and y-axis with the quantity and the unit using the same convention as the column headings of the table (quantity/unit). You can simply copy the headings from the table you have drawn for collecting data. If the data you are plotting is very large (or very small) include the powers of 10 in your axis labels. For example, a graph of cross-sectional area, A, could have the label $A/10^{-7}\,\text{m}^2$.

It is really important that you choose a scale that allows you to see the trend in the data and allows you to easily plot points and read off values.

- You should always use graph paper to draw a graph. Give at least each 2 cm square marked on the graph paper a value. These values should increase in units of 1, 2 or 5. This way, each 2 mm square on the graph has a value that is easy to interpret.
- The plotted points must occupy at least half the grid in both the x and y directions. To achieve this, you can use a false origin. Figure 3.1 shows data plotted using an origin of (0, 0) and using a false origin of (15, 30). The trend in the data is much easier to identify in Figure 3.1b.

a)
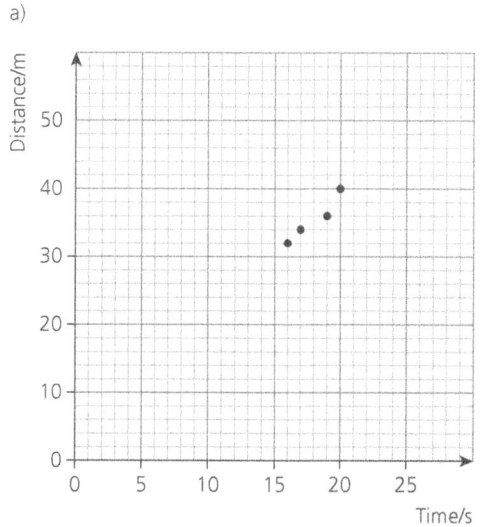

b)
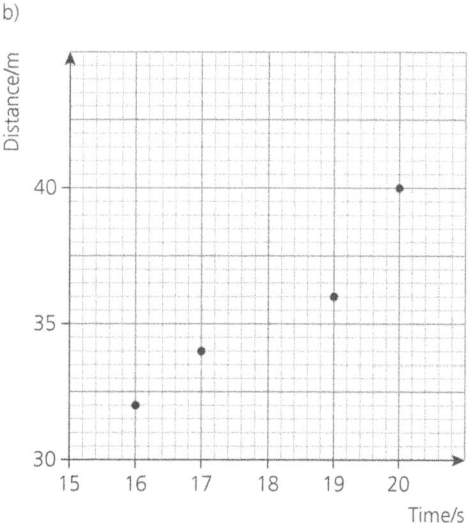

Figure 3.1a The points are clustered in one corner of the graph; **Figure 3.1b** Using a false origin shows the data more clearly

AS LEVEL

3.2.2 Plotting points

The points must be visible. Use a sharp pencil to plot the points and draw either a small cross or an encircled dot. It is very important that the diameter of each point is less than 1 mm.

The points must be plotted accurately. When you look at the data on the graph, if a point seems to be anomalous (does not fit the pattern) then check it again. This could be an incorrectly plotted point or an incorrect measurement. It is best practice to repeat anomalous readings.

3.2.3 Drawing the trend line

The trend line allows you to see if there is a relationship between the variables you have plotted. When you draw the trend line there should be an even distribution of points either side of the line. The line may not go through all the points.

If you suspect a straight-line relationship, use a transparent 30 cm ruler and a sharp pencil. The clear ruler allows you to see all the points. The 30 cm ruler allows you to draw the line in one go so that there are not any bumps.

If you suspect a curved trend line you will need to draw a smooth curve. Turn the graph paper and find a position that allows you to draw the curve in one fluid motion. You can hold the pencil above the paper to practise drawing the shape a few times but it is important to draw the curve in one go. You may need to draw a tangent at a point on the curve to determine the gradient at that point.

If one point does not fit the pattern, then circle it and label is as anomalous. Then, ignore the point when you draw your trend line, whether for a straight or curved trend line.

EXAMPLE 3.3

Plot a graph to show the relationship between the period, T, squared and the length, l, of a pendulum using the data shown in the table:

l/m	T^2/s^2
0.25	0.4
0.50	2.0
0.75	3.2
1.00	4.0
1.50	5.8
2.00	8.0

The completed graph is shown below:

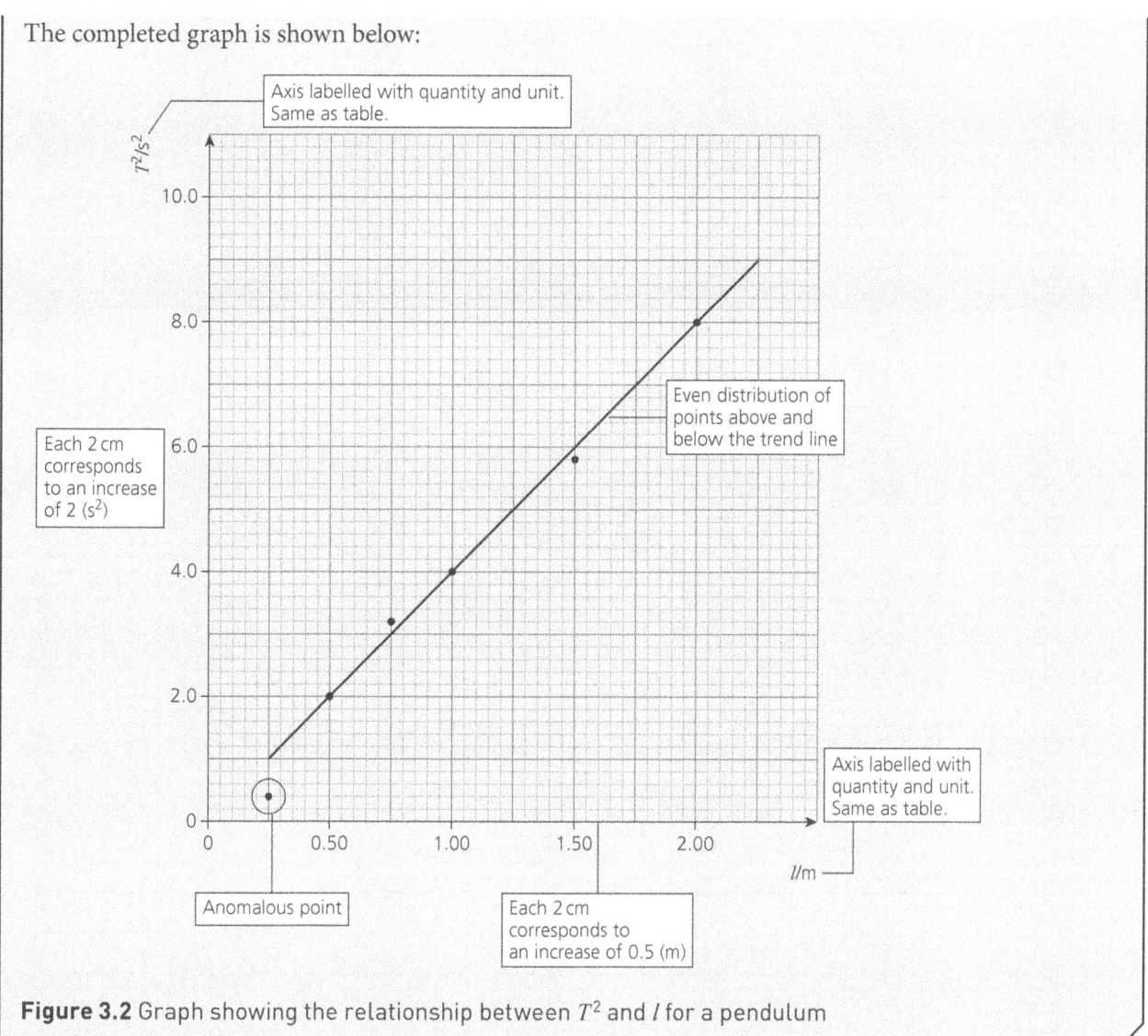

Figure 3.2 Graph showing the relationship between T^2 and l for a pendulum

EXERCISE 3C

Plot the data in the table on the axes below. [5]

M/g	v^2/m² s⁻²
300	2500
350	2760
400	3250
450	3720
500	4160

AS LEVEL

AS LEVEL 4

Analysis, conclusions and evaluation

4.1 Interpreting graphs

Presenting data as a graph helps you to describe the relationship between the variables.

It is important to be able to read data from your trend line graph to the nearest mm (often half the smallest square on the graph paper). Choose coordinates that lie on the gridlines of the graph. It will make reading the values much easier.

4.1.1 Using the equation $y = mx + c$

In the case of a straight-line graph, you can determine the relationship using the equation for a straight-line graph

$$y = mx + c$$

where y is the variable plotted on the y-axis

x is the variable plotted on the x-axis

m is the gradient

c is the intercept on the y-axis (value of y when $x = 0$)

Some of the physics equations you have met can easily be shown to be in the form of a straight-line graph equation, for example, the equation for an object moving with constant acceleration: $v = u + at$.

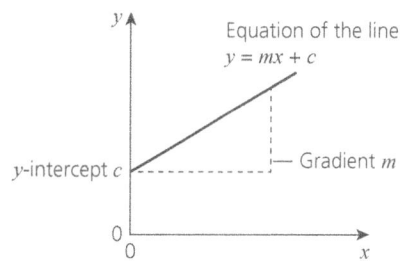

Figure 4.1 The equation for a straight-line graph

If you plot a graph with velocity, v, on the y-axis and time, t, on the x-axis you will get a straight-line graph where the gradient, m, is equal to the acceleration, a, and the y-intercept, c, is equal to the initial velocity, u.

$$v = u + at$$
$$y = mx + c$$

EXAMPLE 4.1

In an investigation, you are told that the relationship between V and I is

$$V = E - Ir$$

where E is the electromotive force of the cell and r is its internal resistance.

To determine E and r you use a variable resistor to take measurements of the potential difference, V, across the resistor and the current, I, in the circuit. You plot a graph with V on the y-axis and I on the x-axis. Sketch the graph and explain how you would use the graph to find E and r.

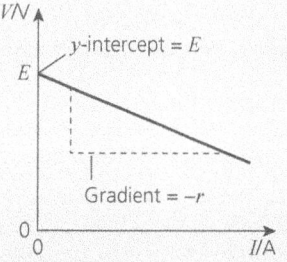

Figure 4.2 Sketch graph for the relationship between V and I

The gradient is equal to $-r$ and the y-intercept is equal to E.

AS LEVEL

EXERCISE 4A

For each relationship and graph, identify the quantity represented by the gradient and the y-intercept.

a Relationship: $y = Ax + B$ Graph: y on the y-axis and x on the x-axis. [2]

....

....

b Relationship: $T^2 = Ax^2 + B$ Graph: T^2 on the y-axis and x^2 on the x-axis. [2]

....

....

c Relationship: $v^2 = u^2 + 2as$ Graph: v^2 on the y-axis and s on the x-axis. [2]

....

....

d Relationship: $R = \dfrac{\rho l}{A}$ Graph: R on the y-axis and l on the x-axis. [2]

....

....

e Relationship: $hf = \phi + eV$ Graph: f on the y-axis and V on the x-axis. [2]

....

....

4.1.2 Calculating the gradient

To calculate the gradient of your trend line you use the equation

$$\text{gradient} = \frac{\Delta y}{\Delta x} = \frac{y_2 - y_1}{x_2 - x_1}$$

Choose coordinates (x_1, y_1) and (x_2, y_2) that lie on the trend line and are at least half the length of the trend line apart. It is much easier if you use points that lie where the grid lines cross.

You should include any powers of 10 included in the units for the x- or y-axis.

If you have a curved trend line, then draw a tangent at a point on the curve. Find the gradient of the tangent. This then gives you the gradient of the curve *at that point*.

> **TIP**
> As in Figure 4.2, draw a clear triangle to show the coordinates you chose and always show your working for your gradient calculation.

To determine the units of the gradient simply divide the units of the quantity on the *y*-axis by the units of the quantity on the *x*-axis. For example, for a graph with T^2/s^2 on the *y*-axis versus l/m on the *x*-axis, the gradient has the following units:

$$\text{gradient} = \frac{s^2}{m} = s^2\,m^{-1}$$

EXAMPLE 4.2

Determine the gradient of the graph shown below at 5 minutes. Give the units.

Method

Draw a tangent at 5 minutes. Find the gradient of the tangent using a large triangle as shown in Figure 4.3.

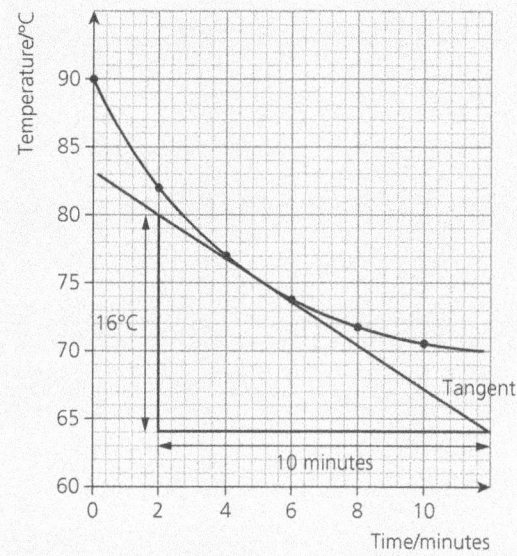

Figure 4.3 Finding the gradient of a tangent

$$\text{gradient} = \frac{\Delta y}{\Delta x} = \frac{80 - 64}{12 - 2} = \frac{16\,°C}{10\,\min} = 1.6\,°C\,\min^{-1}$$

4.1.3 Finding the *y*-intercept

In the equation for a straight-line graph, the *y*-intercept is the value of the *y*-coordinate when $x = 0$. If you have drawn a graph with an origin (0, 0), then you can simply read the value directly from the graph.

If your graph does not have an origin, then read the coordinates from a point *on* your trend line. Use these coordinates and the gradient of your trend line to determine the *y*-intercept. Substitute these values into the equation for a straight line.

$y = mx + c$

$c = y - mx$

EXERCISE 4B

a Determine the gradient and the *y*-intercept of the trend line shown in Figure 4.4. [3]

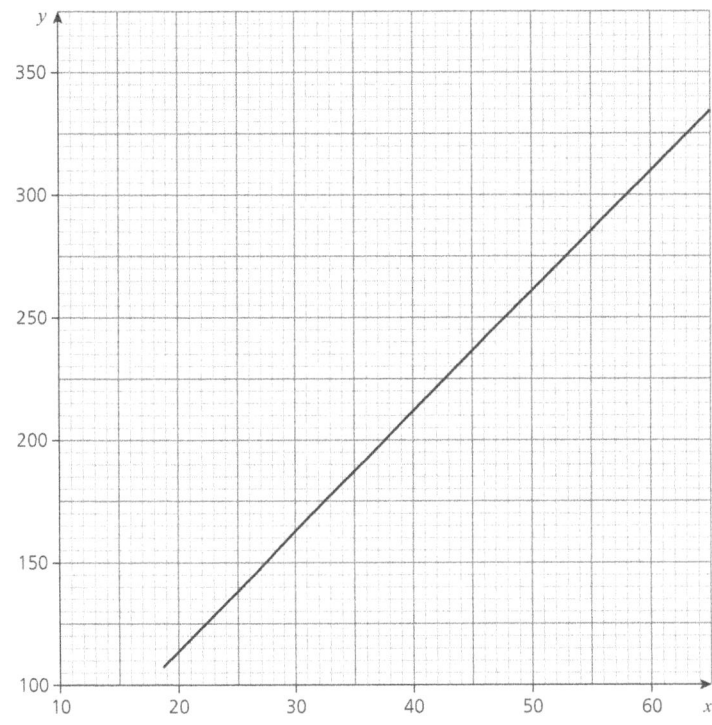

Figure 4.4

b Figure 4.5 shows the graph of R on the x-axis versus $1/I$ on the y-axis.

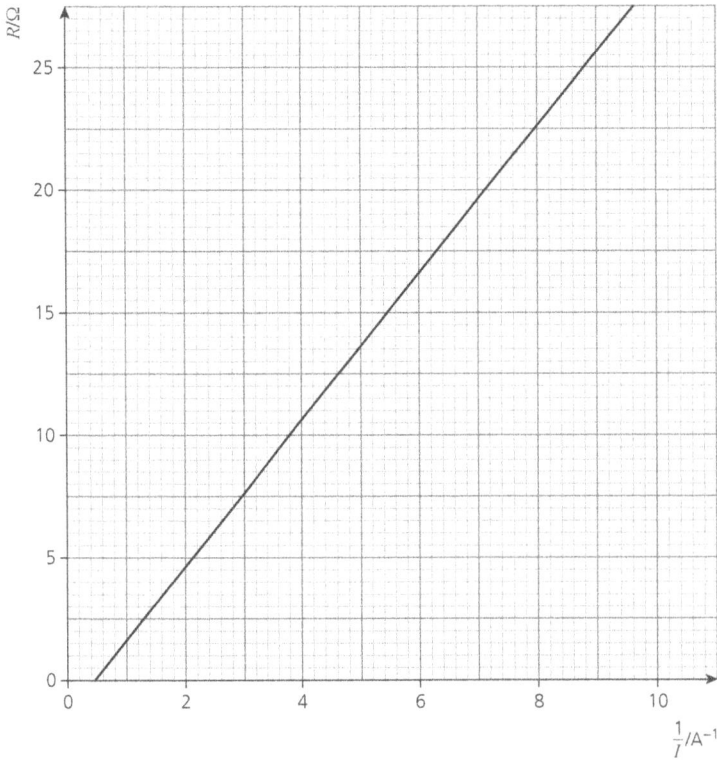

Figure 4.5

i Determine the gradient and the y-intercept for this graph. [3]

....

....

....

ii It is suggested that the quantities I and R are related by the equation
$R = \dfrac{A}{I} - B$

Use your answers to part **i** to give the values of the constants A and B. Give their units. [3]

....

....

4.2 Estimating uncertainties

In Chapter 2, you saw how there is an uncertainty in any measurement due to the smallest interval the instrument can measure. For example, if you use a metre ruler, then you know that the length is measured to the nearest millimetre, for example, 23.2 cm ± 0.1 cm (or 0.323 ± 0.001 m). This is the absolute uncertainty. The uncertainty is expressed in the same units as the measurement.

However, the uncertainty in a measurement also depends on how difficult the measurement was to make or on how much judgement was needed. For example, if you used a metre ruler to measure the height of a bouncing ball, you could not claim to know the height to the nearest millimetre. You are attempting to take the reading while the ball is moving. One way to find the uncertainty is to use the range of your repeated readings. The uncertainty is given by half the range.

$$\text{uncertainty} = \pm \frac{(\text{maximum value} - \text{minimum value})}{2}$$

Remember that the absolute uncertainty should be given to 1 significant figure and the quantity must be quoted to the number of decimal places allowed by the uncertainty.

EXAMPLE 4.3

A student measures the height of bounce as 55.2 cm, 54.5 cm, 55.0 cm, 54.3 cm. Calculate the average height of the bounce and its absolute uncertainty.

Method

Uncertainty $= \pm \dfrac{(55.2 - 54.3)}{2} = 0.45 = 0.5\,\text{cm}$

Mean height of bounce $= 54.75 = 54.8 \pm 0.5\,\text{cm}$

TIP

Sometimes you may have to estimate the absolute uncertainty based on the difficulty of reading the value or level of judgement needed. For example, in the bouncing ball experiment after taking your readings you might estimate the absolute uncertainty as ± 4 mm. Any value greater than 2 mm and less than 6 mm would be an acceptable estimate.

AS LEVEL

You can refer to uncertainties as an absolute value or as a percentage. In Example 4.3, the mean height of bounce is 54.8 ± 0.5 cm, which is 54.8 ± 0.91% as a percentage uncertainty. You can quote percentage uncertainties to 2 significant figures.

> **TIP**
> When calculating percentage uncertainties, the quantity and the uncertainty must be in the same units.

EXERCISE 4C

A student uses a micrometer to measure the diameter of a wire along its length. Their results are 0.24 mm, 0.22 mm, 0.30 mm, 0.26 mm.

Calculate the mean diameter and its absolute uncertainty. [2]

4.3 Combining uncertainties

You might be asked to carry out an investigation to determine a physical quantity such as resistivity or the Young modulus. To find a value, you may make a number of measurements and each of these measurements will have an uncertainty. To find the uncertainty in your final quantity you will have to combine these uncertainties.

- If the quantities are multiplied or divided, **add** their **percentage** uncertainties.
- If the quantities are added or subtracted, **add** their **absolute** uncertainties.
- If the quantity is raised to a power, then **multiply** the uncertainty by that **power**.

EXAMPLE 4.4

A student runs 20 m ± 0.1 m in 4.0 s ± 0.2 s. Calculate the percentage and absolute uncertainty in the speed.

Method

Speed = $\frac{20}{4.0}$ = 5.0 m s^{-1}

% uncertainty in distance = $\frac{0.1}{20} \times 100$ = 0.5%

% uncertainty in time = $\frac{0.2}{4.0} \times 100$ = 5%

Total % uncertainty in speed = 0.5% + 5% = 5.5%

Speed = 5.0 m s^{-1} ± 5.5%

Speed = 5.0 ± 0.3 m s^{-1}

4 Analysis, conclusions and evaluation

EXAMPLE 4.5

A student measures the temperature at the start of an experiment as $20.4 \pm 0.1\,°C$. Thirty seconds later they record a temperature of $25.6 \pm 0.2\,°C$. Calculate the temperature difference giving both absolute and percentage values.

Method

Absolute uncertainty = $0.1\,°C + 0.2\,°C = 0.3\,°C$

Temperature difference = $5.2 \pm 0.3\,°C$

Temperature difference = $5.2 \pm 5.8\,\%$

EXAMPLE 4.6

The diameter of a wire is 0.26 ± 0.01 mm. Calculate the cross-sectional area, A, of the wire and its absolute uncertainty.

Method

% uncertainty in diameter = $\frac{0.01}{0.26} \times 100 = 3.8\,\%$

% uncertainty in $A = 2 \times 3.8 = 7.6\,\%$

$A = \frac{\pi d^2}{4} = \frac{\pi \times 0.26^2}{4} = 0.053\,09\,\text{mm}^2 = 0.053 \pm 0.004\,\text{mm}^2$

TIP

Remember that the cross-sectional area, A, cannot be quoted to more decimal places than the absolute uncertainty. They must be consistent.

EXERCISE 4D

a At a junction in a circuit, current $I_1 = 2.4 \pm 0.1$ A and current $I_2 = 5.5 \pm 0.2$ A meet. Calculate the combined current and the absolute uncertainty in the value. [2]

....

....

b A student takes measurements to determine the resistivity of conducting putty.

$R = 20\,\Omega \pm 0.5\%$ $A = 3.1 \times 10^{-4}\,\text{m}^2 \pm 12\%$ $l = 0.06\,\text{m} \pm 1.7\%$

Calculate the resistivity of the putty and its absolute uncertainty using the equation $\rho = \frac{RA}{l}$. [4]

....

....

AS LEVEL

c The potential difference across a resistor is 5.25 ± 0.07 V and the current is 0.21 ± 0.02 A. Calculate the resistance of the resistor and the absolute uncertainty using the equation $R = \frac{V}{I}$. [4]

d The diameter of a ball bearing is 2.50 ± 0.01 mm. Calculate the volume in mm³ and the absolute uncertainty using the equation $V = \frac{4}{3}\pi r^3$. [4]

e The time to fall a distance of 5.0 ± 0.1 m is 0.9 ± 0.2 s. Calculate g and its absolute uncertainty using the equation $g = \frac{2s}{t^2}$. [5]

4.4 Drawing conclusions

Conclusions state any trend or pattern shown by the data. For example, a straight-line graph through the origin suggests a directly proportional relationship between the variables.

Often you will use the data to determine a constant. For example, a graph of stress versus strain has a gradient that is equal to the Young modulus of the material.

You will want to consider how precise and accurate your data is.
- If repeated values are close to the mean, i.e. if the range is small, then your data is precise. If the points plotted on a graph lie close to your trend line, then your data is precise.
- If the value is close to the true value, then your data is accurate.

4 Analysis, conclusions and evaluation

> **TIP**
> It is possible to have results that are very precise but not very accurate. For example, you could determine a number of values of g in an experiment and have an answer of $11.2 \pm 0.2 \, \text{m s}^{-2}$.
>
> You can also be imprecise and accurate, for example, $g = 9.8 \pm 1.2 \, \text{m s}^{-2}$.

As part of your conclusion, you should consider whether the data collected supports either your prediction or a given relationship between the variables. To do this, calculate the percentage difference between different values of the constant and compare with the experimental uncertainty.

You can calculate the percentage difference as follows:
- If the constant has a known value such as g, the acceleration of free fall, then this is the expected or true value. Compare your value with the true value:

$$\text{percentage difference} = \frac{\text{true value} - \text{experimental value}}{\text{true value}} \times 100\%$$

- If you have limited results and have calculated the value using these, for example, if you have found the resistance of two different lengths of the same wire and used them to find two values for the resistivity of the wire using $\rho = \frac{RA}{l}$, then:

$$\text{percentage difference} = \frac{\text{difference in the two values}}{\text{one of the values}} \times 100\%$$

or

$$\text{percentage difference} = \frac{\text{difference in the two values}}{\text{average of the values}} \times 100\%$$

EXAMPLE 4.7

A student investigated how the period, T, for an oscillating mass, m, on a spring is related to the size of the mass. The table shows their results.

Mass/kg	0.200	0.400
Time/s	0.73	1.10

They were given the relationship
$$T^2 = \frac{m}{b}$$

a Calculate their two values for constant b.

b The experimental uncertainty is 5%. Explain whether the results support the relationship.

Method

a Value 1: $b = \frac{m}{T^2} = \frac{0.200}{0.73^2} = 0.38$

  Value 2: $b = \frac{m}{T^2} = \frac{0.400}{1.10^2} = 0.33$

b Percentage difference $= \frac{\text{difference in the two values}}{\text{average of the values}} \times 100\%$

  Percentage difference $= \frac{0.38 - 0.33}{0.36} \times 100\% = 14\%$

AS LEVEL

or

$$\text{Percentage difference} = \frac{\text{difference in the two values}}{\text{one of the values}} \times 100\%$$

$$\text{Percentage difference} = \frac{0.38 - 0.33}{0.38} \times 100\% = 13\%$$

or

$$\text{Percentage difference} = \frac{0.38 - 0.33}{0.33} \times 100\% = 15\%$$

As the experimental uncertainty is 5% these results do not support the relationship.

TIP

You only have to find the percentage difference by one method.

The values support the relationship if the percentage difference is less than the experimental uncertainty, for example, if the experimental uncertainty had been 18%, the result would have supported the relationship.

As part of a conclusion, you may also make predictions about data outside of your data set. For example, you can collect data on how the pressure of a fixed volume of gas varies with temperature and use it to determine absolute zero (see Figure 4.6).

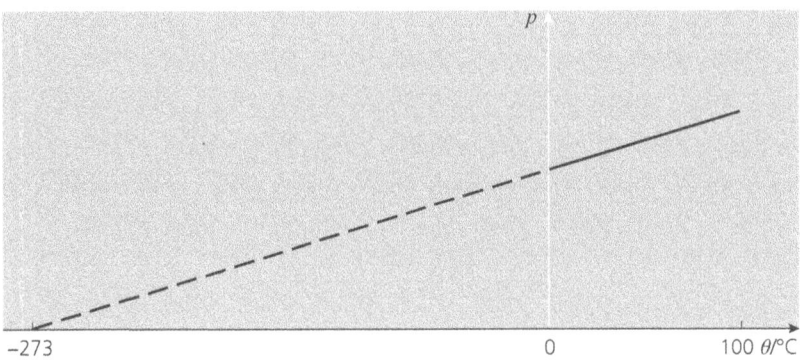

Figure 4.6 The line has been extrapolated (continued) to where $y = 0$; this is absolute zero

EXERCISE 4E

a A student collects data to determine the resistivity of a wire.

Cross sectional area of wire = $2.04 \times 10^{-7}\,\text{m}^2 \pm 4\%$

Length = $0.800 \pm 0.001\,\text{m}$

Resistance = $4.8 \pm 0.1\,\Omega$

i Calculate the resistivity in $\Omega\,\text{m}$ using the equation $\rho = \dfrac{RA}{l}$ [1]

....

....

ii Calculate the percentage uncertainty in the value for resistivity, ρ. [2]

....

....

iii The true value for the resistivity is given as $1.1 \times 10^{-6}\,\Omega\,\text{m}$. Explain whether the experimental data supports this. [2]

....

....

....

4.5 Identifying limitations and suggesting improvements

When you carry out an investigation it is good practice to evaluate your method and the data you collect. To do this, you need to consider the sources of uncertainty in your measuring instruments and in your procedures. This will give you an understanding of how reliable your data is and how you might improve the method if you were to repeat the investigation.

As you complete an experiment, note down any difficulties you encounter as you make your measurements and how this could be improved. If you can improve it with the equipment available, then do so. If not, then note down your ideas to include in your evaluation.

In your evaluation, you can then identify the limitations in each measurement and suggest improvements for that measurement.

EXAMPLE 4.8

You are asked to conduct an experiment to determine the value of the acceleration of free fall, g, using a free fall method. This is the method used:

1. Release the ball from 0.50 m and time how long it takes to hit the floor.
2. Repeat and find an average time.
3. Use the equation $s = ut + \frac{1}{2}at^2$ to determine g.

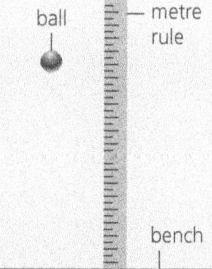

Figure 4.7 Free fall method to determine g

AS LEVEL

To evaluate this method, start by considering the measurements and then any other points you can think of.

- **Measuring s**

  Limitation: The ruler may not be held vertically for each measurement, so the drop height is different.

  Improvement: Hold the ruler in a clamp stand and use a set square to ensure that it is vertical.

  Limitation: It is hard to judge the release height because s is measured to the middle of the ball and it is difficult to judge whether it is in the correct position.

  Improvement: Use a fixed marker to show the exact height where the top of the ball should be placed for the middle to be at 0.50 m. You will have to measure the diameter of the ball using calipers to determine this.

- **Measuring t**

  Limitation: It is difficult to start the stopwatch and release the ball at the same time. It is also difficult to judge when to stop the stopwatch. The time measured is very small, so there is a large percentage uncertainty in the time.

  Improvement: To remove human reaction time error use an electronic timer that is triggered when the ball is released and stopped when the ball hits a metal plate. Or use two light gates to time the ball over a set distance, s, and use the first light gate to determine the initial speed u. Alternatively, film the motion of the ball with a stopwatch in view and then play back the film to measure the time.

- **General points**

  Limitation: In this experiment you are only using one height to determine g. This means that you have one reading repeated many times. This is not enough to draw a valid conclusion.

  Improvement: Collect a series of readings of t for different s values. Plot a graph with t^2 on the y-axis and s on the x-axis. You can then use the trend line to determine g. The gradient of this graph would be equal to $\frac{2}{g}$.

  Limitation: The ball will bounce on the floor and may roll away and become a trip hazard.

  Improvement: Set up so that the ball falls into a box.

TIP

There are a number of suggested improvements given in the example, but you only need to suggest one for each limitation. It may help to draw a diagram to explain your improvements. For example, Figure 4.8 shows the apparatus diagram for an improvement using an electronic timer to time the ball.

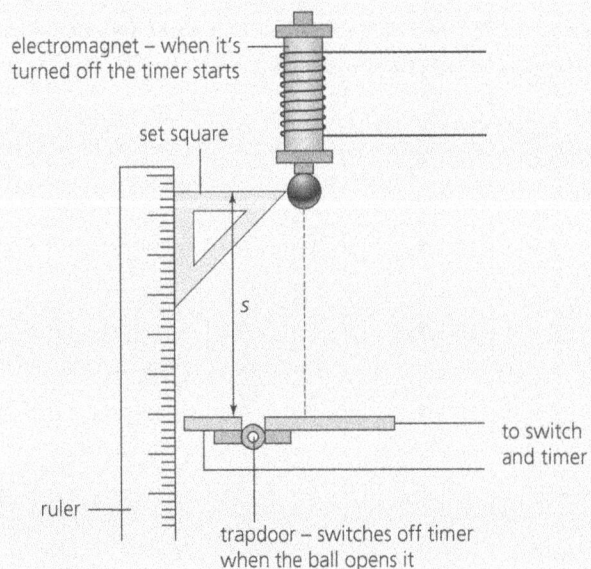

Figure 4.8 Using an electronic timer

4 Analysis, conclusions and evaluation

> **TIP**
> It is important you talk about limitations *specific* to the investigation you have done and then use your knowledge and experience to suggest improvements. For example, normally to reduce the percentage uncertainty you would suggest increasing the time by increasing the height of the drop. However, the ball may reach its terminal speed and no longer be accelerating. This will not improve this experiment.

EXERCISE 4F

A student measures the change in amplitude of a pendulum for each oscillation. They set up the apparatus as shown in Figure 4.9. The period of the pendulum is 1.5 s.

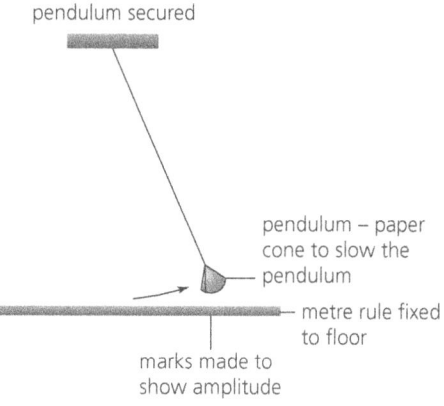

Figure 4.9 Measuring the amplitude of each oscillation of a pendulum

a Describe two sources of uncertainty or limitations of the procedure in this experiment. [2]

1 ....

....

2 ....

....

b Describe two improvements to the experiment. You may suggest alternative equipment. [2]

1 ....

....

2 ....

....

....

AS LEVEL

5 Practice questions

In the practical paper, the questions may include theory not covered in your syllabus. Any theory that you need to know to answer the questions will be supplied – remember that the paper is only assessing your practical skills. Make sure that you only conduct experiments with your teacher's supervision.

> **TIP**
> When you evaluate your method, you have to identify sources of uncertainty or limitations in the method. As you carry out the practical, note down the problems you encounter when taking the measurements.

1 If your teacher and laboratory assistants are able to prepare this practical experiment, according to the Confidential Instructions sent to schools (9702 Paper 32 Q1 May/June 2019), then carry out the experiment. Try to keep to the maximum time allocated, which is 1 hour.

If this is not possible, read through the question and imagine how you would set up the apparatus and take the readings. Then:

- describe how you would measure the length x of wire between P and A and the length y of wire between R and B in the space for collecting results in part **c**.

- use the following data to answer question part **d** onwards.

x/cm	y/cm
20.0	5.3
30.0	8.4
40.0	11.4
50.0	14.0
60.0	17.5
70.0	20.5
80.0	24.0

In this experiment, you will investigate an electrical circuit.

a – Assemble the circuit as shown in Figure 1.1.

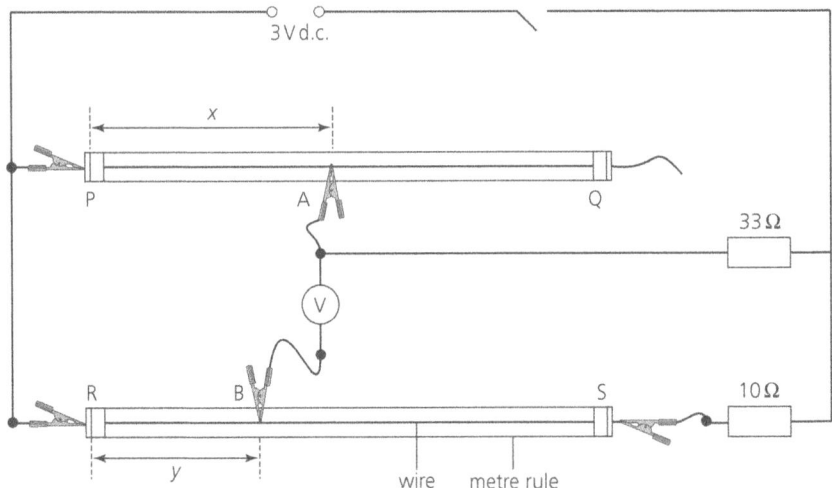

Figure 1.1

- A and B are crocodile clips. PQ and RS are wires. Connect A near the midpoint of PQ.
- Measure and record the length x of wire between P and A.

   $x = $ .... cm [1]

b - Connect B to RS.
- Close the switch.
- Adjust the position of B until the voltmeter reading is as close as possible to zero.
- Measure and record the length y of wire between R and B.

   $y = $ .... cm
- Open the switch. [1]

c Change x and repeat **b** until you have six sets of values of x and y. Record your results in a table. Include values of $\frac{1}{x}$ and $\frac{1}{y}$ in your table. [10]

d i Plot a graph of $\frac{1}{y}$ on the y-axis against $\frac{1}{x}$ on the x-axis. [3]

  ii Draw the straight line of best fit. [1]

  iii Determine the gradient and y-intercept of this line.

   gradient = ...

   y-intercept = ... [2]

AS LEVEL

e It is suggested that the quantities y and x are related by the equation

$$\frac{1}{y} = \frac{a}{x} + b$$

where a and b are constants.

Use your answers in **diii** to determine the values of a and b.

Give appropriate units.

$a = $ ....

$b = $ .... [2]

[Total: 20]

Adapted from Cambridge International AS & A Level Physics 9702 Paper 32 Q1 May/June 2019

2 If your teacher and laboratory assistants are able to prepare this practical experiment, according to the Confidential Instructions sent to schools (9702 Paper 34 Q1 May/June 2021), then carry out the experiment. Try to keep to the maximum time allocated, which is 1 hour.

If this is not possible, read through the question and imagine how you would set up the apparatus and take the readings. Then:

- describe how you would measure distance C and time the oscillations accurately in the space for collecting results in part **b**
- use the following data to answer question part **c** onwards.

C/cm	15.0	20.0	25.0	30.0	35.0	40.0
T/s	0.70	0.79	0.88	0.95	1.02	1.07

a In this experiment, you will investigate the oscillations of a chain.
- Assemble the apparatus as shown in Figure 2.1 with each nail held securely in a boss and at the same height above the bench. Position the stands so that the distance between the nails is approximately 60 cm.

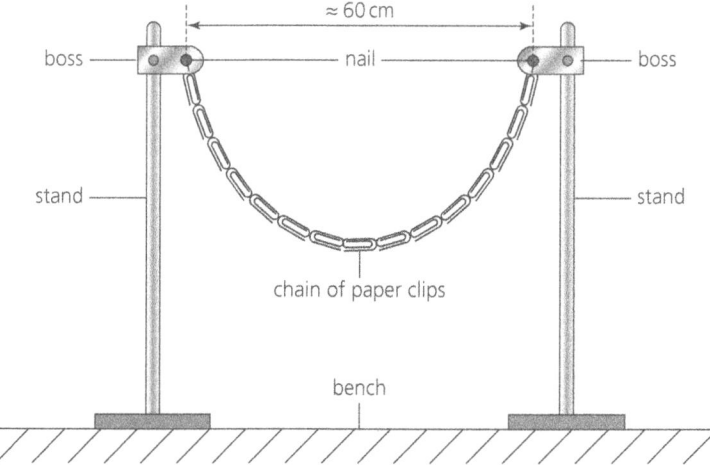

Figure 2.1

- Rest one of the metre rules on the nails, as shown in Figure 2.2.

AS LEVEL

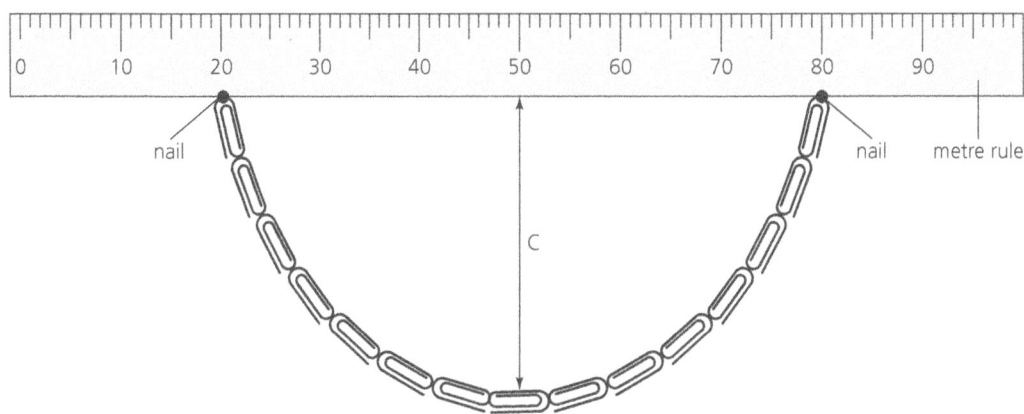

Figure 2.2

- The vertical distance between the horizontal metre rule and the lowest part of the chain is C.

   Using the other metre rule, measure and record C.

   i $C =$ .... cm [1]

   ii – Push the bottom of the chain a short distance away from you. Release it so that it swings towards and away from you.

   – Take measurements to determine the period T of these oscillations.

   $T =$ .... [2]

b Repeat **a** with different distances between the stands until you have six sets of values of C and T.

All values of C must be greater than 15 cm.

Record your results in a table. Include values of $\frac{1}{T}$ and $\frac{1}{\sqrt{C}}$ in your table. [9]

c i Plot a graph of $\frac{1}{T}$ on the y-axis against $\frac{1}{\sqrt{C}}$ on the x-axis. [3]

   ii Draw the straight line of best fit. [1]

   iii Determine the gradient and y-intercept of this line.

   gradient = ...

   y-intercept = ... [2]

5 Practice questions

AS LEVEL

d It is suggested that the quantities T and C are related by the equation

$$\frac{1}{T} = \frac{a}{\sqrt{C}} + b$$

where a and b are constants.

Use your answers in **ciii** to determine the values of a and b.

Give appropriate units.

$a = $ ....

$b = $ .... [2]

[Total: 20]

Adapted from Cambridge International AS & A Level Physics 9702 Paper 34 Q1 May/June 2021

3 If your teacher and laboratory assistants are able to prepare this practical experiment, according to the Confidential Instructions sent to schools (9702 Paper 34 Q1 October/November 2020), carry out the experiment under exam conditions.

In this experiment, you will investigate the equilibrium of a metre rule with a chain attached.
- Attach the boss to the stand at a height of approximately 60 cm above the bench.
- Assemble the apparatus as shown in Figure 3.1 with the nail held securely in the boss.
- Attach one end of the chain of paper clips to the string loop and allow the other end of the chain to rest on the bench.
- Attach the piece of adhesive putty to the metre rule approximately 40 cm from the nail.

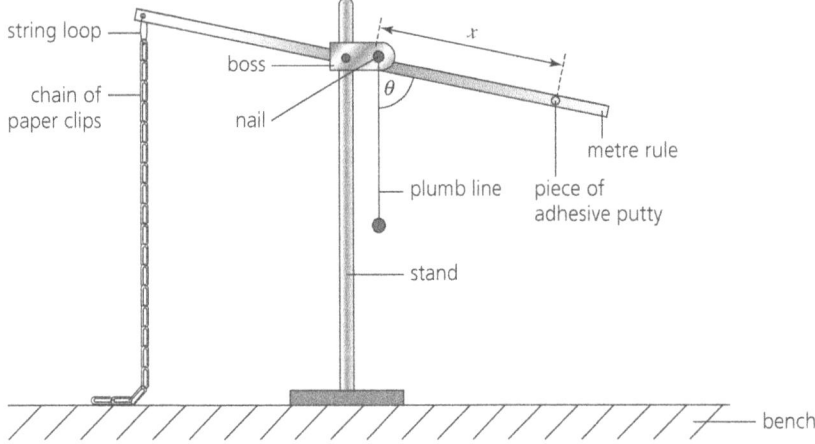

Figure 3.1

- Measure and record the distance x between the nail and the centre of the piece of adhesive putty, as shown in Figure 3.1.

a $x = $ .... cm [1]

- Measure and record the angle θ between the metre rule and the plumb line, as shown in Figure 3.1.

b $\theta = $ ....° [1]

c Vary x and measure θ until you have six sets of values of x and θ.

Do not use values of x less than 15 cm.

Record your results in a table. Include values of $\cos\theta$ in your table. [10]

d i Plot a graph of $\cos\theta$ on the y-axis against x on the x-axis. [3]

ii Draw the straight line of best fit. [1]

iii Determine the gradient and y-intercept of this line.

gradient = ...

y-intercept = ... [2]

AS LEVEL

e It is suggested that the quantities θ and x are related by the equation

$$\cos\theta = ax + b$$

where a and b are constants.

Use your answers in **diii** to determine the values of a and b.

Give appropriate units.

$a = $ ....

$b = $ .... [2]

[Total: 20]

Adapted from Cambridge International AS & A Level Physics 9702 Paper 34 Q1 October/November 2020

4 If your teacher and laboratory assistants are able to prepare this practical experiment, according to the Confidential Instructions sent to schools (9702 Paper 34 Q2 May/June 2020), then carry out the experiment. Try to keep to the maximum time allocated, which is 1 hour.

If this is not possible, then read through the question and imagine how you would set up the apparatus and take the readings.

- When a question asks for a measurement, describe how you would take that measurement.
- For part **ai**, use a value of 0.69 s for T.
- Use the following data to answer question part **e** onwards.

y/m	0.100	0.050
H/m	0.235	0.180

In this experiment, you will investigate the amplitude of oscillations of a mass suspended from a spring.

a i - Assemble the apparatus as shown in Figure 4.1.
- Pull the mass hanger and slotted masses down through a short distance. Release them so that they oscillate vertically.
- Measure and record the period T of the oscillations.

$T = $ .... s   [1]

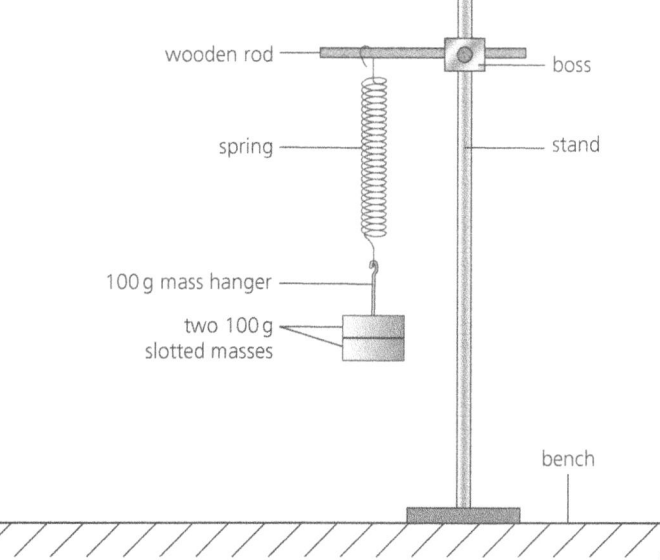

Figure 4.1

AS LEVEL

ii Calculate the spring constant, k, using

$$k = \frac{4\pi^2 M}{T^2}$$

where $M = 0.300$ kg.

$k = $ .... N m^{-1} [1]

b – Slide the two 100 g slotted masses to the top of the mass hanger as shown in Figure 4.2.

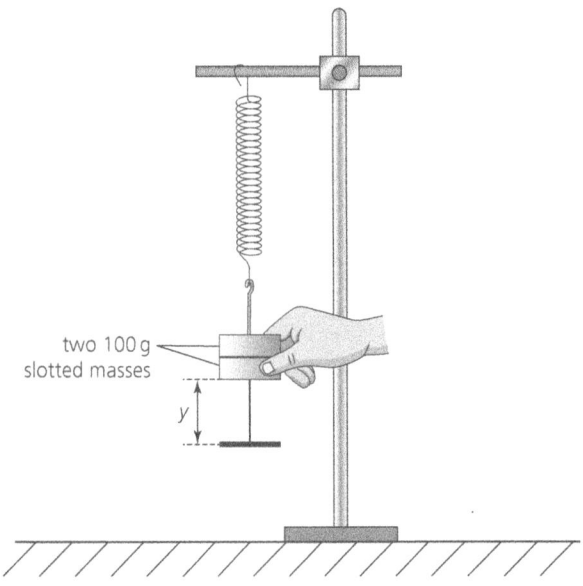

Figure 4.2

– The height of the slotted masses above the base of the mass hanger is y, as shown in Figure 4.2.

Measure and record y.

$y = $ .... m [1]

c – Drop the two 100 g slotted masses. The masses and the mass hanger will oscillate vertically as shown in Figure 4.3.

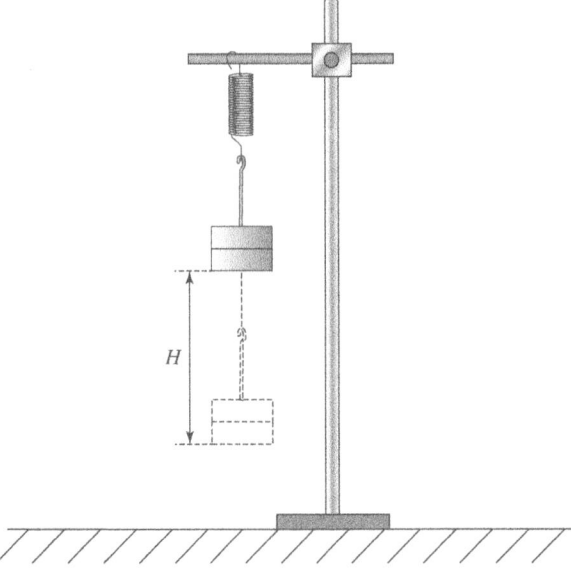

Figure 4.3

- The distance between the lowest and highest positions of the oscillating mass hanger is H as shown in Figure 4.3. Measure and record H.

$H = \text{..............................}$ m [2]

d Estimate the percentage uncertainty in your value of H. Show your working.

percentage uncertainty = ... [1]

e Repeat **b** and **c** but this time sliding the two slotted masses approximately half-way up the mass hanger.

$y = \text{..............................}$ m

$H = \text{..............................}$ m [2]

f It is suggested that the relationship between H and y is

$$H = c\sqrt{y}$$

where c is a constant.

i Using your data, calculate two values of c.

first value of $c = $ ....

second value of $c = $ .... [1]

ii Justify the number of significant figures you have given for your values of c. [1]

....

....

iii Explain whether your results for **fi** support the suggested relationship. [1]

....

....

AS LEVEL

g Theory suggests that an approximate value for the acceleration of free fall g is given by

$$g = \frac{c^2 k}{8m}$$

where $m = 0.200$ kg.

Use your value of k from **aii** and your first value of c to calculate g. Include an appropriate unit.

$g =$ .... [1]

h i Describe four sources of uncertainty or limitations of the procedure for this experiment.

1 ....

....

2 ....

....

3 ....

....

4 ....

....

[4]

ii Describe four improvements that could be made to this experiment. You may suggest the use of other apparatus or different procedures.

1 ....

....

2 ....

....

3 ....

....

4 ....

....

[4]

Total [20]

Adapted from Cambridge International AS & A Level Physics 9702 Paper 34 Q2 May/June 2020

5 Practice questions

5 If your teacher and laboratory assistants are able to prepare this practical experiment, according to the Confidential Instructions sent to schools (9702 Paper 34 Q2 October/November 2019), then carry out the experiment. Try to keep to the maximum time allocated, which is 1 hour.

If this is not possible, then read through the question and imagine how you would set up the apparatus and take the readings.

- When a question asks for a measurement, describe how you would take that measurement.
- For each measurement, sample data has been provided here: $h = 9$ mm. *1st measurements:* $W = 2.0$ N, $r = 22$ mm, $F = 2.7 \pm 0.2$ N. *2nd measurements:* $W = 1.0$ N, $r = 18$ mm, $F = 1.7 \pm 0.2$ N. Use this data to answer the questions as expected.

In this experiment, you will investigate the force needed to pull a cylinder up a step.

a Measure the thickness h of the board, as shown in Figure 5.1.

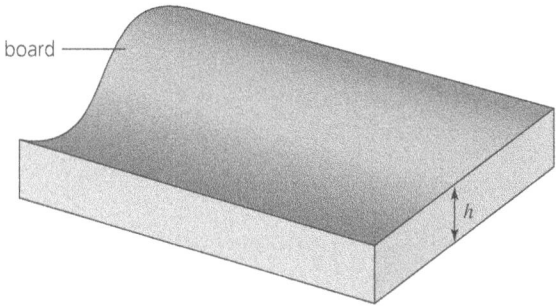

Figure 5.1

$h = $ .... [1]

b Suspend the two larger (100 g) slotted masses from the newton meter using the loop of thread, as shown in Figure 5.2.

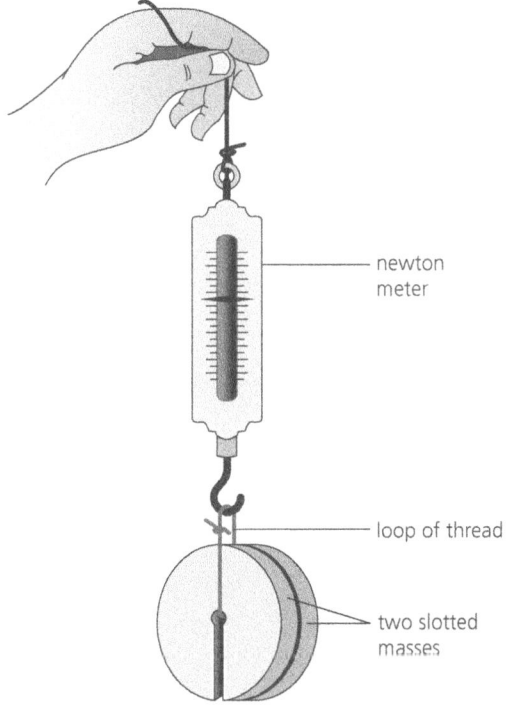

Figure 5.2

Record the total weight W of these masses.

$W = $ .... [1]

AS LEVEL

c i Take measurements to find the radius r of one of the larger slotted masses.

$r = $ .... [1]

ii The value of α is given by

$$\sin \alpha = \frac{(r - h)}{r}$$

Calculate α.

$\alpha = $ .... ° [1]

iii Justify the number of significant figures you have given for your value of α. [1]

....

....

d Place the board on the bench to make a step. Stand the two larger slotted masses on their edges next to the step with their slots at the top, as shown in Figure 5.3.

Attach the loop of thread to the masses and the newton meter, as shown in Figure 5.3.

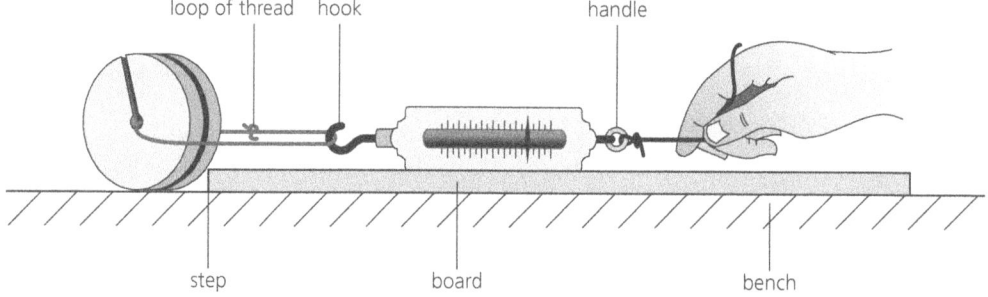

Figure 5.3

Pull the handle of the newton meter horizontally and at right angles to the step. The force required to just start the slotted masses rolling up the step is F. Measure and record F.

$F = $ .... N [2]

e Estimate the percentage uncertainty in your value of F.

percentage uncertainty = ... [1]

f Repeat **b**, **ci**, **cii** and **d** using the two smaller 50 g slotted masses.

W = ... N

r = ... mm

α = ...°

F = ... N [2]

g It is suggested that the relationship between F, W and α is

$$F = \frac{kW}{\tan \alpha}$$

where k is a constant.

i Using your data, calculate two values of k.

first value of k = ...

second value of k = ... [1]

AS LEVEL

ii Explain whether your results support the suggested relationship. [1]

h i Describe four sources of uncertainty or limitations of the procedure for this experiment.

1 ....

2 ....

3 ....

4 ....

[4]

ii Describe four improvements that could be made to this experiment. You may suggest the use of other apparatus or different procedures.

1 ....

2 ....

3 ....

4 ....

[4]

[Total: 20]

Adapted from Cambridge International AS & A Level Physics 9702 Paper 34 Q2 October/November 2019

5 Practice questions

6 If your teacher and laboratory assistants are able to prepare this practical experiment, according to the Confidential Instructions sent to schools (9702 Paper 33 Q2 October/November 2021), then carry out the experiment under exam conditions.

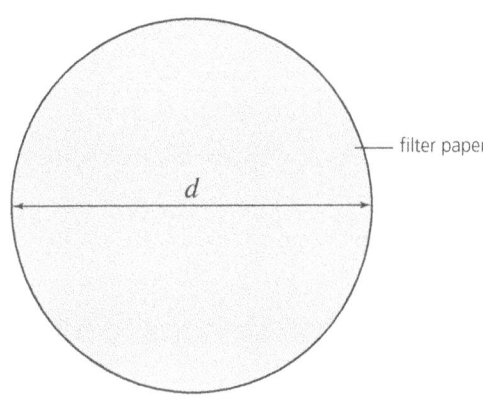

Figure 6.1

In this experiment, you will investigate the time taken for filter papers to fall in air.
- You have been provided with filter papers of two different sizes.
  Take one sheet of the smaller filter paper.
- The diameter of one sheet of filter paper is d, as shown in Figure 6.1.

a i Measure and record d.

   $d = $ .... cm [2]

ii Calculate the area A of the filter paper using

$$A = \frac{\pi d^2}{4}$$

   $A = $ .... cm² [1]

iii Justify your number of significant figures that you have given for your value of A. [1]

....

....

b i Set up the apparatus as shown in Figure 6.2.
- Hold the six sheets of the smaller filter paper at the top of the metre rule, as shown in Figure 6.2. Release the filter papers and start the stopwatch.
- The time between release and the filter papers hitting the bench is t.

Measure and record t.

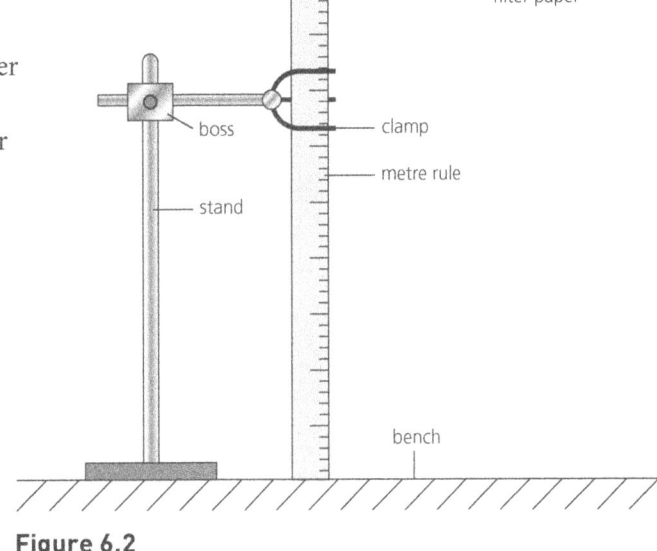

Figure 6.2

   $t = $ .... s [2]

ii Estimate the percentage uncertainty in t. Show your working.

   percentage uncertainty = ... [1]

iii Measure and record the total mass m of the sheets of smaller filter paper.

   $m = $ .... [1]

AS LEVEL

c i Repeat **ai** and **aii** using one of the larger sheets of filter paper.

   $d =$ .... cm

   $A =$ .... cm² [1]

   ii Using two sheets of the larger filter paper, repeat **bi** and **biii**.

   $t =$ .... s

   $m =$ .... [1]

d It is suggested that the relationship between t, m and A is

   $kt = mA$

   where k is a constant.

   i Using your data, calculate two values of k.

   first value of $k =$ ....

   second value of $k =$ .... [1]

   ii Explain whether your results support the suggested relationship.

   ....

   ....

   [1]

e i Describe four sources of uncertainty or limitations of the procedure for this experiment.

   1 ....

   ....

   2 ....

   ....

   3 ....

   ....

   4 ....

   ....

   [4]

  ii Describe four improvements that could be made to this experiment. You may suggest the use of other apparatus or different procedures.

   1 ....

   ....

   2 ....

   ....

   3 ....

   ....

   4 ....

   ....

   [4]

   [Total: 20]

Adapted from Cambridge International AS & A Level Physics 9702 Paper 33 Q2 October/November 2021

A LEVEL

6 Planning

During your physics course, you have learnt how to use many different measuring instruments and types of laboratory equipment. When you design an investigation to test a relationship, you need to use and apply all that knowledge.

There are four steps to writing a plan:
- defining the problem
- writing a method
- explaining how you will analyse the data
- considering safety.

6.1 Defining the problem

The first step in designing an investigation is to determine the purpose of the investigation. For example, are you trying to investigate a relationship between two variables or determine a constant, or perhaps both?

Start with the hypothesis and identify the independent, dependent and control variables.
- The independent variable is the one that you will change.
- The dependent variable is the one that will be measured as a result of changing the independent variable.
- Control variables are the variables which could also affect the outcome of the experiment. These are the variables that will need to be kept constant. It is important to include in the method how this will be done.

EXAMPLE 6.1

A student supports a beam at both ends and suspends a mass, m, from the middle. They wish to investigate how the sag, s, depends on the mass, m, suspended. It is suggested that the sag, s, varies with m according to the relationship

$$s = \frac{mgL^3}{4bd^3E}$$

where L is the length, b is the width, d is the depth of the beam and E is the Young modulus.

The independent, dependent and control variables for this experiment are as follows.
- Independent variable is m.
- Dependent variable is s.
- Control variables are L, b, d and E.

TIP

g (gravitational field strength) is a constant but it is not a control variable (unless you are planning to complete the experiment on another planet). The value of g only varies within 0.5% across the planet.

EXERCISE 6A

a A student investigates the relationship between the diameter, d, of a metal wire and its resistance, R. They are given the equation

$$R = \frac{4\rho L}{\pi d^2}$$

where ρ is the resistivity of the metal and L is the length of the wire. Identify the independent, dependent and control variables. [3]

....

....

....

b A student investigates how the extension, x, of a metal wire varies with length, L, of the wire. They suspend a mass, m, from the wire. They are given the equation

$$E = \frac{mgL}{Ax}$$

where E is the Young modulus of the wire, and A is the cross-sectional area of the wire. Identify the independent, dependent and control variables. [3]

....

....

....

6.2 Methods of data collection

Once you have identified your variables you have to describe how to carry out the investigation. Your plan must be clear enough so that another person could follow the instructions and accurately collect data.

In your method you must describe how you will vary the independent variable, how you will measure the dependent and independent variables and how you will keep the control variables constant. Details of how you will ensure that the data is as precise and accurate as possible must also be included.

It is often useful to start with a clear, labelled diagram. The diagram can include how the measurements are to be made.

A LEVEL

EXAMPLE 6.2

A student supports a beam at both ends and suspends a mass, m, from the middle. They wish to investigate how the sag, s, depends on the mass, m, suspended.

It is suggested that the sag, s, varies with m according to the relationship

$$s = \frac{mgL^3}{4bd^3E}$$

where L is the length, b is the width, d is the depth of the beam and E is the Young modulus.

Write a method describing how to take the measurements needed to investigate how s varies with m.

Method

Set up the apparatus as shown in Figure 6.1.

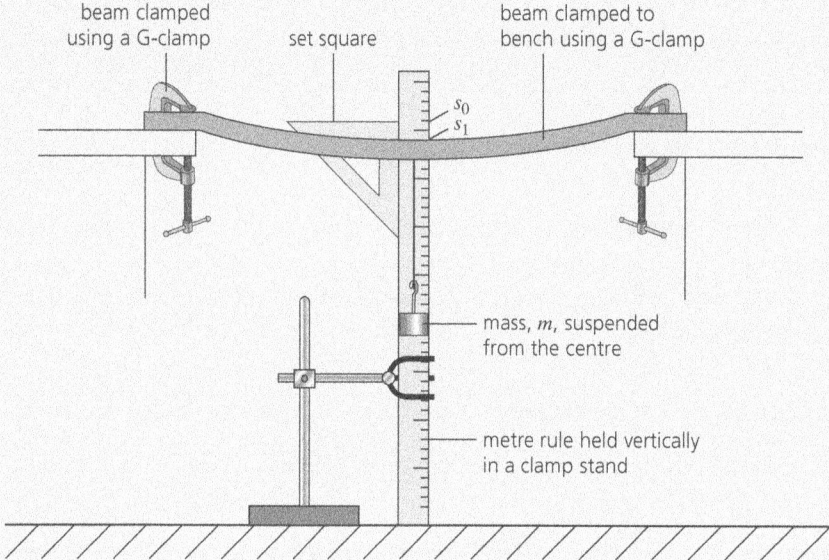

Figure 6.1 Apparatus to investigate how s varies with m

The independent variable is the mass, m.

1. Accurately measure the mass using a balance.

2. Measure the width and depth of the beam using calipers. The beam may not be uniform along its length, so to improve accuracy, measure both width and depth along the length of the beam and find mean values. Use the same beam throughout the investigation.

3. Clamp the beam to the benches as shown in the diagram. Use a spirit level to ensure that it is horizontal.

4. Measure the length on both sides to ensure that it is perpendicular to the bench and to accurately identify the centre point of the beam.

5. To ensure the measurement of s is accurate place a vertical metre ruler held in place by a clamp stand. Use a set square to ensure that the ruler is vertical and perpendicular to the beam when it is not loaded.

6. Record the reading on the ruler: this is s_0.

The mass is suspended from the centre of the beam by a string tied to the middle.

7. Read the new reading on the ruler (s_1). The sag is the difference between s_0 and s_1.

8. Remove the mass and then add it on again to get a repeat reading of the sag.

9. Determine the average sag.

10. Repeat the method with increasing m.

> **TIP**
> Notice that the method describes how the control variables are measured and kept constant. This is important. It also includes details of how to improve the accuracy of the measurements such as fixing the metre rule vertically and using a set square to ensure that it is vertical.

EXERCISE 6B

A student investigates the relationship between the diameter, d, of a metal wire and its resistance, R. They are given the equation

$$R = \frac{4\rho L}{\pi d^2}$$

where ρ is the resistivity of the metal and L is the length of the wire. Write a method describing how to take the measurements needed to investigate how R varies with d. [8]

A LEVEL

Chapter 2 of this guide looked at how to choose and use appropriate measuring instruments. However, at A level you may need to use more technical equipment to measure your values, for example, using an oscilloscope or a data logger or even filming an experiment.

6.2.1 Using an oscilloscope

There are many different types of oscilloscope; Figure 6.2 shows one type you may have seen. All oscilloscopes have a screen which is overlaid with a grid of squares usually 1 cm × 1 cm. You connect your source to the *y*-input and adjust the volt/division and time base until a suitable trace is obtained.

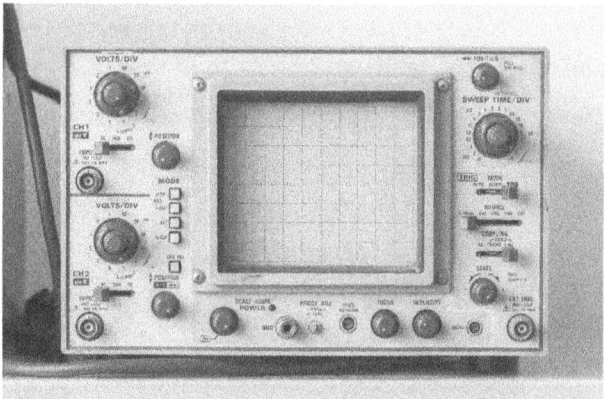

Figure 6.2 An oscilloscope

Each horizontal square represents a certain time interval. This is determined by the dial marked time base or time/div. For example, if this is set to $1\,\text{ms}\,\text{cm}^{-1}$, then each horizontal square represents 1 ms.

To determine the time period, you measure the distance for one complete cycle. This distance is then multiplied by the setting on the time base. The frequency is then equal to $\frac{1}{T}$.

Each vertical square represents a certain voltage interval. The dial marked volts/div or *y*-gain determines this. For example, if it is set to $0.2\,\text{V}\,\text{cm}^{-1}$, then each vertical square represents 0.2 V. The peak voltage is determined using the amplitude of the trace. This value is then multiplied by the setting on the volts/div. To determine the peak current, you will need to know the resistance R of the component you are investigating. Then use the equation $I = \frac{V}{R}$.

To improve the accuracy of your readings from an oscilloscope, set the time/div and volts/div so that the wave fills the screen. You can move the wave form vertically and horizontally so that it lines up with the grid for ease of measurement.

EXAMPLE 6.3

A *y*-input is connected across a component connected to an alternating voltage source. The resistance of the component is 100 Ω. Determine the frequency of the alternating source and the peak current.

Time base is set to $0.5\,\text{ms}\,\text{cm}^{-1}$

The volts/div is set to $0.2\,\text{V}\,\text{cm}^{-1}$

Method

Time period = $6.0\,\text{cm} \times 0.5 \times 10^{-3} = 3.0 \times 10^{-3}\,\text{s}$

Frequency = $f = \frac{1}{T} = \frac{1}{3.0 \times 10^{-3}} = 330\,\text{Hz}$

Peak voltage = $1.4\,\text{cm} \times 0.2\,\text{V}\,\text{cm}^{-1} = 0.28\,\text{V}$

Peak current = $I = \frac{V}{R} = \frac{0.28}{100} = 2.8\,\text{mA}$

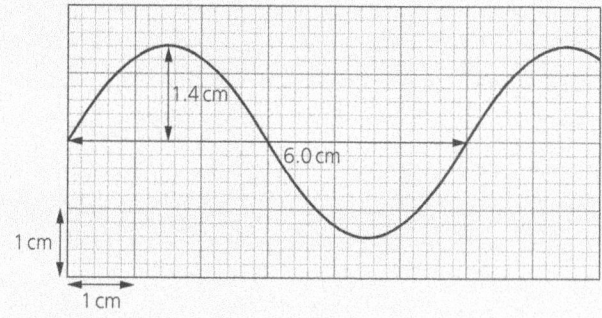

Figure 6.3 An alternating voltage

6 Planning

EXERCISE 6C

Determine the frequency and peak voltage from the oscilloscope trace shown in Figure 6.4.

Time base is set to 10 ms cm^{-1}
The volts/div is set to 5.0 V cm^{-1} [3]

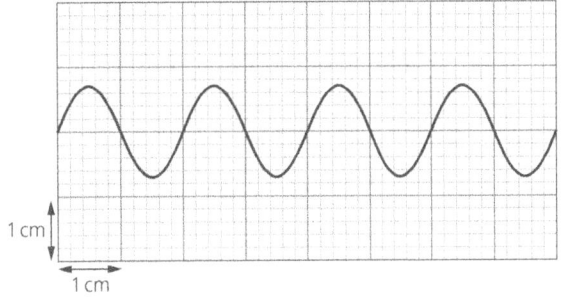

Figure 6.4 An oscilloscope trace

6.2.2 Using a data logger

Data loggers can be used with different sensor inputs.

Light gates

Light gates are used to time events accurately. As they are triggered automatically there is no human error and the uncertainty is 1×10^{-5} s. Light gates have an infrared transmitter and receiver. As the object passes through the light gate, it blocks the infrared beam and starts the counter. The counter stops when the beam is detected again at the receiver.

- Determining velocity. If you know the length (s) of the object blocking the light gate you can use the equation $v = \frac{s}{t}$ to determine the velocity.
- Determining acceleration. The data logger can be programmed to record the time between each triggering of a light gate. This means you can determine acceleration using one light gate and a

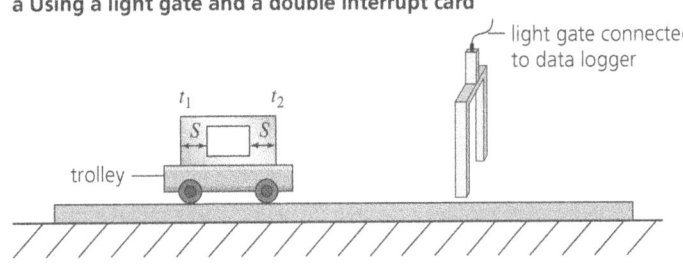

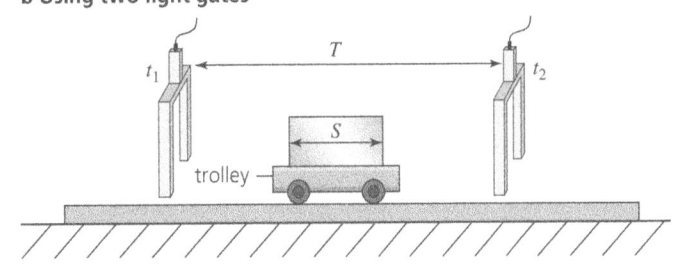

Figure 6.5 Using light gates to determine acceleration

'double interrupt' card or using two light gates. The initial velocity u $\left(u = \frac{s}{t_1}\right)$ and final velocity v $\left(v = \frac{s}{t_2}\right)$ can be calculated and the time taken (T), which allows you to determine acceleration $\left(a = \frac{v-u}{T}\right)$.

Other sensors

Other sensors such as for temperature, voltage or current can be connected to a data logger. The advantages of using a data logger when measuring these variables is that you can set the time interval between readings. This means you can either collect data over a long period such as every 24 hours or a lot of data over a short interval of time. For example, if you are monitoring the potential difference or current as a capacitor discharges it is very difficult to record the potential difference or current at the same time intervals when the values are rapidly changing. This leads to human error. To improve accuracy of measurement, the data logger can be programmed to collect measurements every second and the data can be plotted directly from the data logger.

A LEVEL

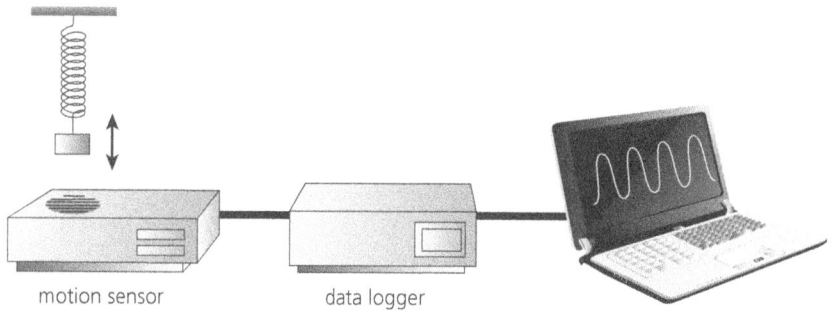

Figure 6.6 Using a motion sensor

For example, a motion sensor sends out ultrasonic pulses which are reflected back to the sensor. The data logger then uses this data to accurately determine the position of the object at a set time (see Figure 6.6). This data can then be used to determine how the velocity and acceleration of the object change with time.

Taking measurements from film

On occasions, measurements have to be made on moving objects, for example when measuring the frequency or wavelength of a ripple tank wave. To improve accuracy, film the wave with a stopwatch and metre ruler in clear view. This allows you to play back the film, pause it and take measurements. You have to be mindful of the position of the camera to avoid parallax errors when reading from an analogue scale such as a ruler.

EXERCISE 6D

The trolley in Figure 6.7 is pulled to the right and released. You are investigating how the amplitude of the oscillations changes over time. The time period is 0.5 s. Describe how you would accurately measure the amplitude of the oscillations. [3]

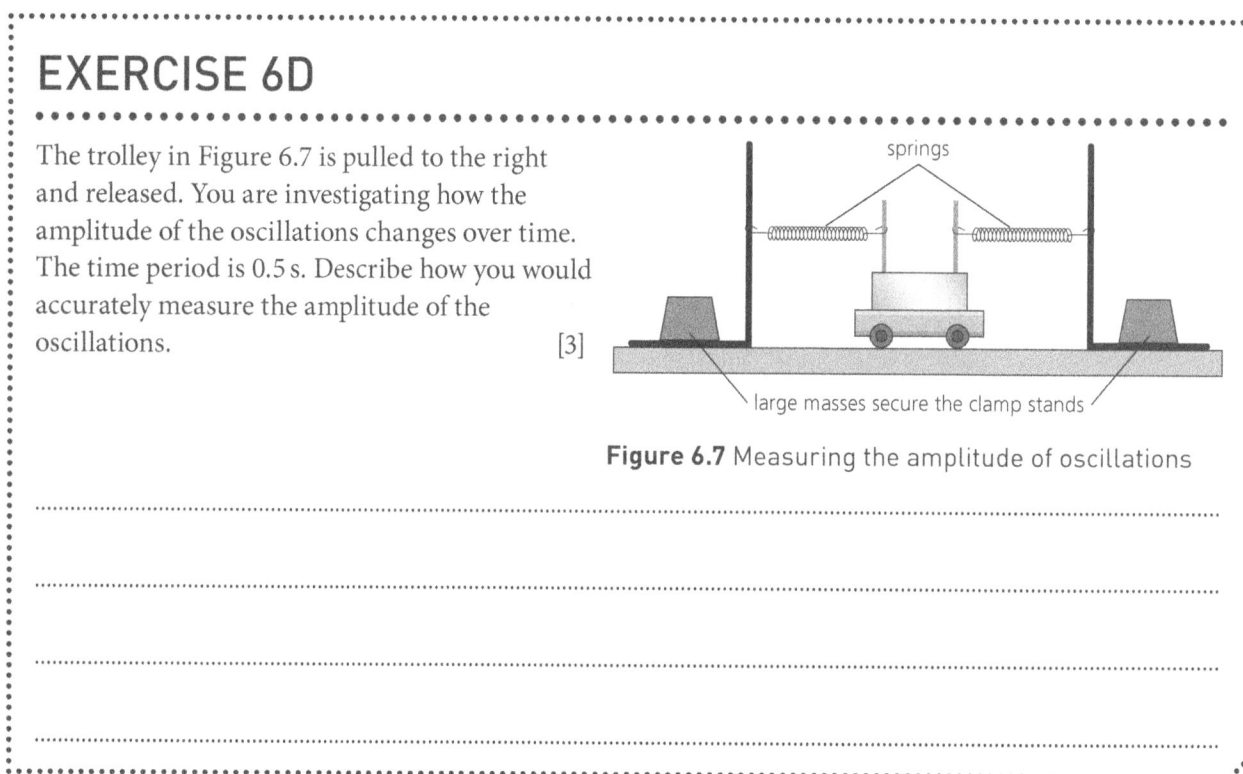

Figure 6.7 Measuring the amplitude of oscillations

....

....

....

....

6.3 Methods of analysis

As part of the method, it is important to outline how you will analyse the data in order to reach a conclusion or determine a constant. This may include calculating values from the raw data.

Usually, you will plot a graph and this should be a straight-line graph. To decide what to plot on the x- and y-axis look at the proposed relationship. For example, see whether you can rearrange the equation to the form $y = mx + c$ and find the constant, m, from the gradient and the constant, c, from the y-intercept.

In your plan, you must include which quantity is plotted on the x-axis and which is plotted on the y-axis. It is important to explain how you will know that the relationship is correct from your graph. For example, for a given experiment, you might be expecting a straight-line graph through the origin or a straight-line graph with a y-intercept. You must also include how you will use the gradient and/or intercept to determine any constants.

In Chapter 7, you will look in more detail at how you can test different types of relationships such as a relationship with an unknown power or an exponential relationship.

EXAMPLE 6.4

A student supports a beam at both ends and suspends a mass, m, from the middle. They wish to investigate how the sag, s, depends on the mass, m, suspended. It is suggested that s varies with m according to the relationship

$$s = \frac{mgL^3}{4bd^3 E}$$

where L is the length, b is the width, d is the depth of the beam and E is the Young modulus.

Explain how you will use your results to prove this relationship and find a value of E.

Method

Plot a graph with s on the y-axis and m on the x-axis. If the relationship is correct, then it will be a straight-line graph through the origin.

The Young modulus can be calculated using the equation

$$E = \frac{gL^3}{4bd^3 \times gradient}$$

EXERCISE 6E

a A student investigates how the length of pendulum affects the period. They are given the relationship

$$T = 2\pi \sqrt{\frac{l}{g}}$$

They vary the length, l, and measure the period, T. Explain how they could use their results to prove the relationship and determine a value of g. [3]

....

....

....

....

b A student investigates how the voltage, V, required to stop electrons leaving a surface varies with the frequency, f, of the incident light. They are given the relationship

$$eV = hf - \phi$$

where e, h and ϕ are constants.

Explain how they could use their results to prove the relationship and determine a value of h and ϕ. [4]

6.4 Considering safety

It is important to consider safety in an activity and in Chapter 1 you looked at how this process is done. In your plan, you must assess the risk and describe any precautions you should take to reduce that risk.

EXAMPLE 6.5

A student supports a beam at both ends and suspends a mass, m, from the middle. They wish to investigate how the sag, s, depends on the mass, m, suspended. It is suggested that s varies with m according to the relationship

$$s = \frac{mgL^3}{4bd^3E}$$

where L is the length, b is the width, d is the depth of the beam and E is the Young modulus.

Assess the risk of this investigation and describe any suitable precautions to take to reduce the risk.

Method

- The mass may fall because the beam snaps or the string breaks. Place a sandbox beneath the experiment to cushion the fall and prevent your feet from being under the mass in case it falls.
- The beam may break and shatter. Wear safety specs to protect your eyes.
- Observe the beam carefully as you add the masses and stop if you see any breaks starting.

TIP

Examples 6.1, 6.2, 6.4 and 6.5 combined provide a model for how to write a plan for an experiment. To help structure your plan, you might want to use these headings:
- Apparatus diagram: *Labelled apparatus diagram clearly showing how measurements are to be taken.*
- Variables: *Identify the independent, dependent and control variables.*
- Method: *Describe how **all** measurements will be made including how the control variables will be kept constant.*
- Safety: *Identify hazards, risk and how you will reduce the risk.*
- Analysis: *Describe the graph to plot and how it will prove the relationship. Where appropriate include details of how you will find the constant.*

EXERCISE 6F

Assess the risks of an investigation into the discharge of a 25 V electrolytic capacitor. [2]

....

....

A LEVEL
7 Analysis, conclusions and evaluation

In this chapter, you will build on the analysis you completed at AS Level covered in Chapter 4. You will look at how to test more complex relationships using logarithms, address uncertainties in results tables and determine uncertainties in values calculated from graphs.

7.1 Data analysis

So far you have focused on relationships where you can rearrange a given equation to the form $y = mx + c$ and find the constant, m, from the gradient and the constant, c, from the y-intercept. In this section, you are going to learn what graphs to plot when dealing with relationships with an unknown power or exponential relationships. This is done using logarithms.

7.1.1 Relationships in the form $A = kB^n$

For relationships in the form $A = kB^n$, where k is a constant, you plot a graph of $\log A$ on the y-axis and $\log B$ on the x-axis.

To understand why this works, take logarithms to base 10 on both sides of the equation.

$$A = kB^n$$

$$\log A = \log k + n \log B$$

> **TIP**
> Remember the rules for logarithms:
> $\log(pq) = \log p + \log q$
> $\log(p^n) = n \log p$

If you plot a graph with $\log A$ on the y-axis and $\log B$ on the x-axis and if the relationship is correct, then you will get a straight-line graph. Compare this with the equation for a straight-line graph $y = mx + c$.

The gradient is equal to the power n.

The y-intercept is equal to $\log k$.

Therefore $k = 10^{y\text{-intercept}}$

> ### EXAMPLE 7.1
>
> A student is investigating the relationship between force, F, required to keep a bung moving in a horizontal circle, and its velocity, v. They think that the relationship is $F = av^2$ but they are not sure. Explain what graph they can plot to prove the relationship and how they would determine the constant a.
>
> **Method**
>
> Plot $\log F$ on the y-axis and $\log v$ on the x-axis. If the student is correct, then the gradient will be 2.
>
> y-intercept $= \log a$
>
> Therefore, $a = 10^{y\text{-intercept}}$

> **TIP**
> If you know that the relationship is $F = av^2$ and plot a graph with F on the y-axis and v^2 on the x-axis, you will get a straight line through the origin and the gradient will be equal to a. Use a log–log graph when you suspect a power relationship but do not know what the power is.

68 Photocopying prohibited

7 Analysis, conclusions and evaluation

7.1.2 Relationships in the form $y = ae^{kx}$

For exponential relationships in the form $y = ae^{kx}$, you plot a graph of $\ln y$ on the y-axis and x on the x-axis.

To understand why this works, take logarithms to the base e (natural logarithms) on both sides of the equation.

$$y = ae^{kx}$$
$$\ln y = \ln a + kx$$

If you plot $\ln y$ on the y-axis and x on the x-axis and this relationship is correct, you will get a straight-line graph. Compare this with the equation for a straight line.

gradient = k

y-intercept = $\ln a$

Therefore, $a = e^{y\text{-intercept}}$

EXAMPLE 7.2

A researcher suspects exponential growth in number, N, for a colony of fruit flies with time, t, of the form

$$N = N_0 e^{kt}$$

Where N_0 is the initial number of fruit flies and k is a constant. They record how N changes with time t. Explain how the researcher could prove this relationship.

Method

Plot a graph with $\ln N$ on the y-axis and t on the x-axis. If they are correct the graph will be a straight line with a y-intercept.

gradient = k

y-intercept = $\ln N_0$

Therefore, $N_0 = e^{y\text{-intercept}}$

EXERCISE 7A

In each case, explain how you would test the relationship and determine the values of the constants.

a A student investigates how the speed, v, of a shallow wave is related to the depth, d. They are given the relationship [3]

$$v^2 = gd$$

....

....

....

A LEVEL

b A student investigates how the current, I, changes with time, t, as a capacitor discharges. They are given the relationship and the value of the resistor, R [4]

$$I = I_0 e^{-\frac{t}{CR}}$$

....

....

....

....

c A student investigates data on how the period of orbit, T, of a planet depends on the radius, r, of the orbit. They are given the relationship [4]

$$T = k r^n$$

....

....

....

....

....

7.2 Tables of results

In the AS Level work covered in Chapter 3, you looked at how data is recorded in tables. Each column in a table should have the quantity and units in the form 'quantity/unit'. All raw data in a column is written to the same number of decimal places. For calculated values in tables the number of significant figures will be the same as (or one more than) the number with the fewest number of significant figures used to calculate it.

For logarithms, the heading should be written as log (quantity/unit). This is because a logarithm does not have a unit. For example, to determine the natural logarithm of the current, I, the heading would be log (I/A); for calculated logarithms you record these to the same number of decimal places as the number of significant figures it was calculated from (or one more than). For example, if $r = 1.56$ m, then log (r/m) = 0.193 to 3 decimal places (or 0.1931 to 4 decimal places).

It is good practice to include the absolute uncertainties next to every value in your table. In Chapter 3, you looked at how you could use half the range of your repeats to give a value for raw data. There are a number of methods for determining the absolute uncertainty in your calculated values. In a table, the maximum and minimum method is the easiest to use. This is covered in Section 7.4 Treatment of uncertainties.

In the table, absolute uncertainties can be quoted to more than one significant figure.

EXERCISE 7B

In each question, calculate the value and give the answer to the correct number of decimal places or significant figures as appropriate. Include the correct units.

a Calculate the natural log of the voltage if $V = 10.5\,\text{V}$. [2]

....

b Calculate the value of T^2x, where $T = 2.5\,\text{s}$ and $x = 0.238\,\text{m}$. [2]

....

c Calculate the logarithm to base 10 of the mass if $m = 0.52\,\text{g}$. [2]

....

7.3 Graphs

The uncertainty in each measurement needs to be shown on the graph. This is done by plotting error bars. The error bar shows the uncertainty in each point plotted.

A trend line called the line of best fit is drawn and this must pass through all the error bars. This is used to calculate your gradient and used to determine the y-intercept.

To determine the uncertainty in your gradient and y-intercept, a second trend line is added called the line of worst fit. This line must still go through all the error bars but is the least acceptable line. The line of worst fit can either be the steepest or shallowest trend line possible.

The lines of best and worst fit must be clearly labelled. Figure 7.1 shows a graph with error bars and lines of worst and best fit labelled. A line drawn perpendicular to each error bar shows the end of the error bar clearly.

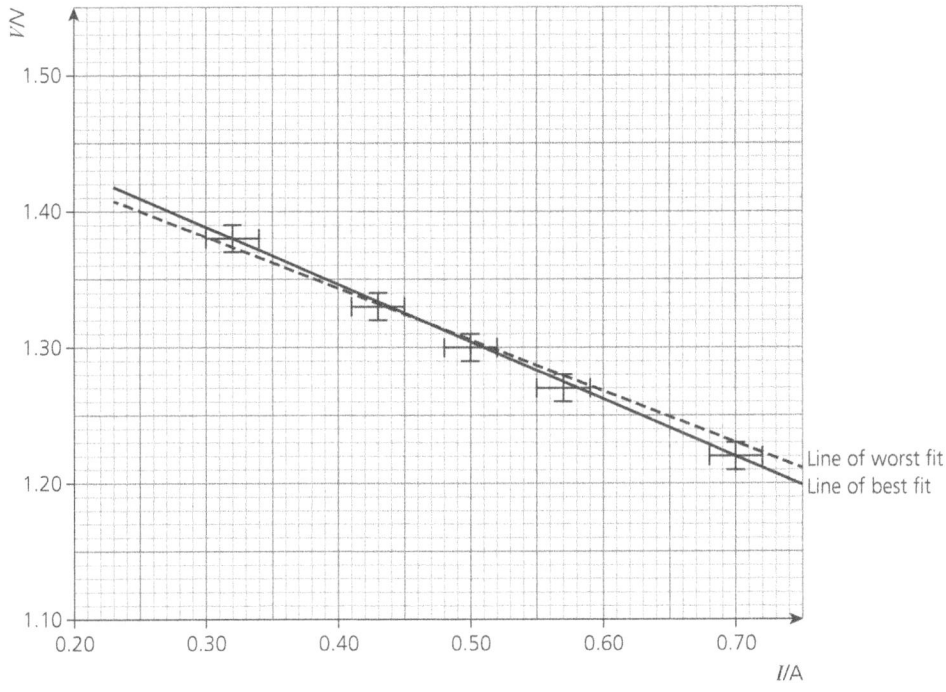

Figure 7.1 Error bars allow you to draw the lines of best fit and worst fit on your graph; for this graph, the line of worst fit is the shallowest line

A LEVEL

7.4 Treatment of uncertainties

In this section, you will look at how to treat uncertainties for all parts of your analysis and conclusion including graphs.

7.4.1 Converting from absolute uncertainties to fractional or percentage and back

The absolute uncertainty is the uncertainty expressed in the same units as the measurement. For example, using a newton meter to measure a force of 2.5 N you may estimate the uncertainty as 0.2 N. This is the absolute uncertainty and is written as 2.5 ± 0.2 N.

The absolute uncertainty can also be written as a fractional uncertainty:

$$\text{fractional uncertainty} = \frac{\Delta x}{x}.$$

For example, for the force measurement above, the fractional uncertainty $= \frac{0.2}{2.5} = 0.08$

It can also be written as a percentage uncertainty:

$$\text{percentage uncertainty} = \frac{\Delta x}{x} \times 100$$

For example, the percentage uncertainty in the force measurement is $\frac{0.2}{2.5} \times 100 = 8\%$

It is important you can also convert percentage or fractional to absolute uncertainties:

$$\text{absolute uncertainty} = \frac{\text{percentage uncertainty}}{100} \times \text{quantity}$$
$$= \text{fractional uncertainty} \times \text{quantity}$$

For example, a resistance is determined to be $48 \pm 7\%$ Ω. The absolute uncertainty is given by

$$\text{absolute uncertainty} = \frac{7}{100} \times 48 = 3.36$$

Resistance $= 48 \pm 3$ Ω

> **TIP**
> In a conclusion, the absolute uncertainty should be expressed to 1 significant figure and the final conclusion must be consistent with it. For example, it would be incorrect to write $V = 2.54 \pm 0.1$ V. The answer should be 2.5 ± 0.1 V.

EXERCISE 7C

a Convert the following absolute uncertainties into fractional and percentage uncertainties.

  i 22.8 ± 0.4 cm [2]

  ii 250 ± 8 mA [2]

b Convert the following to absolute uncertainties.

  i $P = 25.67 \text{ W} \pm 5\%$ [2]

  ....

  ....

  ii $a = 15.29 \text{ ms}^{-2} \pm 0.025$ [2]

  ....

  ....

7.4.2 Calculating uncertainties for derived values

Most uncertainties can be combined using the following simple rules.
- If the quantities are multiplied or divided, then add their percentage or fractional uncertainties.
- If the quantities are added or subtracted, then add their absolute uncertainties.
- If the quantity is raised to a power, then multiply the fractional or percentage uncertainty by that power.

Look back at Chapter 4 if you need to revise this.

To find the uncertainty in the logarithm of a number or in the sine of an angle, you can use the maximum and minimum method of determining the uncertainty in a derived value.

For example, if you have measured V as $8.2 \text{ V} \pm 0.5 \text{ V}$ and wish to determine the uncertainty in natural logarithm of V, $\ln V = 2.10$

$V_{max} = 8.7 \text{ V} \quad \ln V_{max} = 2.16$

$V_{min} = 7.7 \text{ V} \quad \ln V_{min} = 2.04$

The absolute uncertainty in V could be given by

  absolute uncertainty $= \ln V_{max} - \ln V = 2.16 - 2.10 = 0.06$

or

  absolute uncertainty $= \ln V - \ln V_{min} = 2.10 - 2.04 = 0.06$

or

  absolute uncertainty $\frac{1}{2}(\ln V_{max} - \ln V_{min}) = \frac{1}{2}(2.16 - 2.04) = 0.06$

> **TIP**
> Any of these methods is acceptable. The first method (difference between maximum and experimental value) is probably the easiest to do when analysing your practical data.

You can use the maximum and minimum method to work out the uncertainty for other quantities rather than combining percentage or fractional uncertainties. It is very useful when working out the uncertainties in calculated values in a table of results.

However, take care if the derived quantity is calculated by dividing one quantity by another. For example, calculating resistance, R, using voltage, V, and current, I.

  maximum value of resistance $= R_{max} = \dfrac{V_{max}}{I_{min}}$

  minimum value of resistance $= R_{min} = \dfrac{V_{min}}{I_{max}}$

A LEVEL

EXAMPLE 7.3

A student investigates how the period, T, of a pendulum varies with length, L. They time five oscillations for different lengths.

It is suggested that $T = \dfrac{2\pi L^q}{A}$

where A and q are constants.

The values of L, $5T$ and $\log L$ are given in this table:

L/m	$5T$/s	T/s	$\log(L$/m$)$	$\log(T$/s$)$
2.20	15.00 ± 0.4	3.00 ±	0.342	
2.00	14.25 ± 0.4	2.85 ±	0.301	
1.80	13.45 ± 0.4	2.69 ±	0.255	
1.60	12.60 ± 0.4	2.52 ±	0.204	
1.40	11.85 ± 0.4	2.37 ±	0.146	
1.20	10.95 ± 0.4	2.19 ±	0.079	

Calculate the values of $\log T$ and include the uncertainties in T and $\log T$. [3]

Method

The uncertainty in T is $\dfrac{\text{uncertainty in } 5T}{5}$

The uncertainty in $\log T$ is $= \log T_{\max} - \log T$

(You could also use $\log T - \log T_{\min}$ or $\frac{1}{2}\left(\log T_{\max} - \log T_{\min}\right)$)

Results are shown in the table below.

L/m	$5T$/s	T/s	$\log(L$/m$)$	$\log(T$/s$)$
2.20	15.00 ± 0.4	3.00 ± 0.08	0.342	0.477 ± 0.012
2.00	14.25 ± 0.4	2.85 ± 0.08	0.301	0.455 ± 0.012
1.80	13.45 ± 0.4	2.69 ± 0.08	0.255	0.430 ± 0.012
1.60	12.60 ± 0.4	2.52 ± 0.08	0.204	0.401 ± 0.014
1.40	11.85 ± 0.4	2.37 ± 0.08	0.146	0.375 ± 0.014
1.20	10.95 ± 0.4	2.19 ± 0.08	0.079	0.340 ± 0.016

TIP

Note the log values can be quoted to 3 or 4 decimal places because both the length and time are known to 3 significant figures.

7 Analysis, conclusions and evaluation

EXERCISE 7D

a A student is determining the extension of a rubber cord. The original length is 9.5 ± 0.2 cm and the new length after a force is applied is 12.0 ± 0.4 cm. Give the value of the extension and the absolute uncertainty in the measurement. [2]

....

b The current, I, is 2.4 ± 0.2 A in a component of resistance, $R = 110\,\Omega \pm 5\%$. Calculate the power and the absolute uncertainty using the equation $P = I^2 R$. [4]

....

....

....

c Calculate the missing values and add uncertainties to the data in these tables.

i [4]

Mass/g	Time for 10 oscillations/s			Time for 1 oscillation/s	log (T/s)	log (m/g)
	1	2	Mean			
$100 \pm 5\%$	2.5	3.4				

ii [2]

T/ms	$\frac{1}{T}$/s^{-1}
4.2 ± 0.2	

iii Value of $X_0 = 15.2 \pm 0.2$ Wm^{-2} [2]

X_1/Wm^{-2}	$\dfrac{X_1}{X_0}$
24.8 ± 0.2	

7.4.3 Calculating uncertainties from graphs

Lines of worst and best fit are used to estimate the uncertainty in the gradient and y-intercept.

The absolute uncertainty in the gradient is given by

   absolute uncertainty = gradient of best fit line − gradient of worst fit line

The absolute uncertainty in the y-intercept is given by

   absolute uncertainty = y-intercept of best fit line − y-intercept of worst fit line

A LEVEL

EXAMPLE 7.4

Figure 7.2 shows the results from Example 7.3 plotted on a graph.

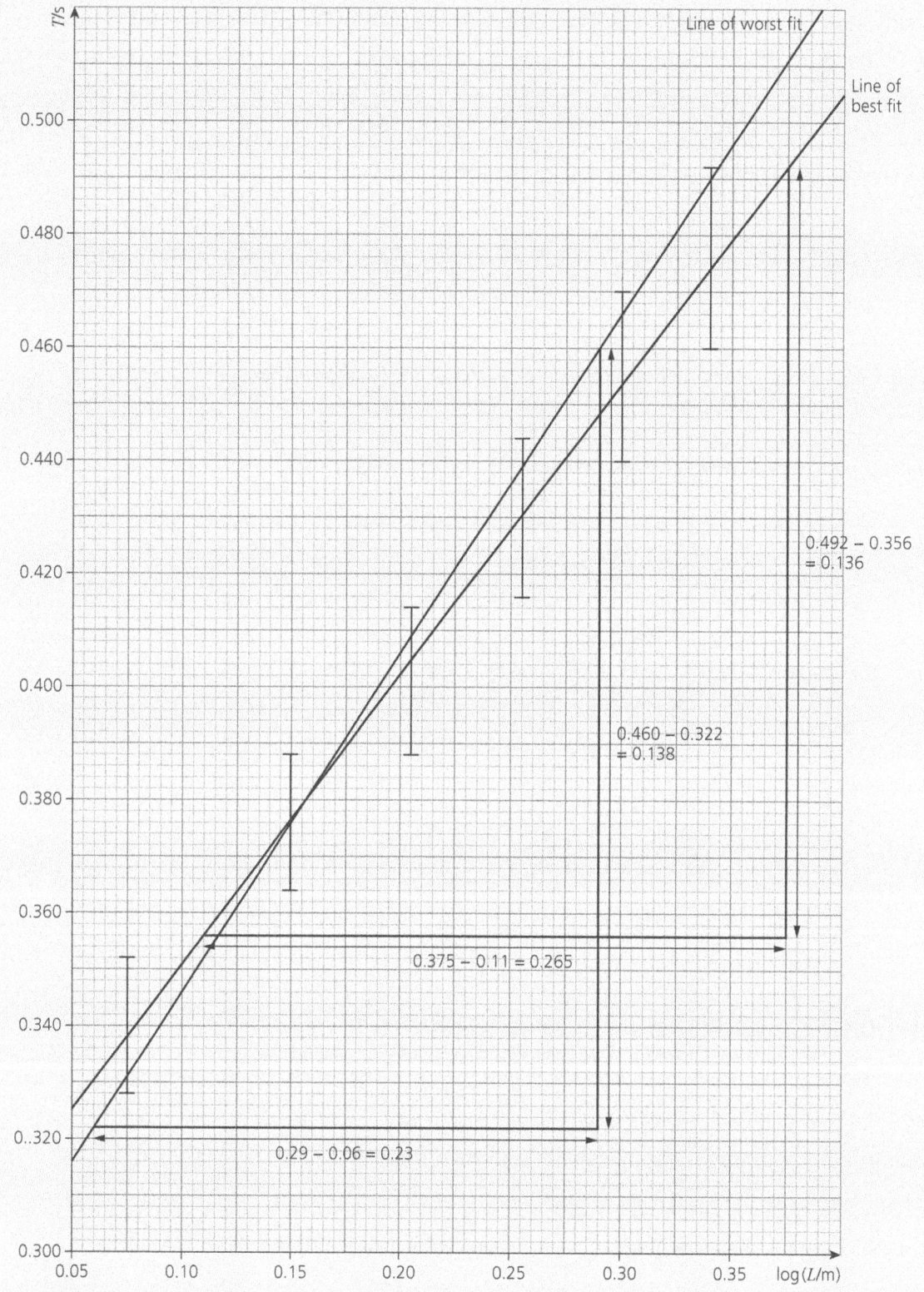

Figure 7.2 Graph showing lines of worst and best fit

a Determine the gradient of the line of best fit and its uncertainty. [2]

b Determine the y-intercept of the line of best fit and its uncertainty. [2]

Method

a Gradient of line of best fit = $\frac{(0.492 - 0.356)}{(0.375 - 0.110)} = 0.513$

Gradient of line of worst fit = $\frac{(0.460 - 0.322)}{(0.29 - 0.06)} = 0.600$

Absolute uncertainty in gradient = difference between gradients

Absolute uncertainty in gradient = $0.600 - 0.513 = 0.09$

Gradient = 0.51 ± 0.09

b y-intercept line of best fit:

$y = mx + c$

Using $m = 0.51$, $y = 0.493$, $x = 0.375$

$0.492 = (0.51 \times 0.375) + c$

$c = 0.3$

y-intercept line of worst fit:

$y = mx + c$

Using $m = 0.60$, $y = 0.460$, $x = 0.29$

$0.460 = (0.60 \times 0.29) + c$

$c = 0.29$

Absolute uncertainty in y-intercept = difference between y-intercepts

Absolute uncertainty in y-intercept = $0.30 - 0.29 = 0.01$

y-intercept = 0.30 ± 0.01

EXERCISE 7E

Figure 7.1 shows lines of best fit and worst fit for data collected during an experiment to determine the emf of a cell. Use the graph in Figure 7.1 to

a determine the gradient of the line of best fit and its uncertainty [4]

b determine the y-intercept of the line of best fit and its uncertainty. [3]

A LEVEL

7.5 Conclusion

In your conclusion you should determine the gradient and y-intercept from your graphs (as shown in Chapter 4). Use these values to determine constants and their units where appropriate. You should also determine their absolute uncertainty.

It is important to show all the steps in your working. State the equation, then substitute in your numbers before calculating.

EXAMPLE 7.5

A student investigates how the period, T, of a pendulum varies with length, L. They time five oscillations for different lengths.

It is suggested that $T = \dfrac{2\pi L^q}{A}$

where A and q are constants.

a Using the gradient and y-intercept calculated in Example 7.4, calculate values for A and q. You do not have to include their absolute uncertainty or their units.

b Using your answer to part **a**, calculate the length, L, needed to give a period of 5.0 s.

Method

a $\log T = \log\left(\dfrac{2\pi}{A}\right) + q \log L$

$\log T$ is plotted on the y-axis and $\log L$ is plotted on the x-axis.

$q = \text{gradient} = 0.51$

$\log\left(\dfrac{2\pi}{A}\right) = y\text{-intercept}$

$\left(\dfrac{2\pi}{A}\right) = 10^{y\text{-intercept}}$

$\left(\dfrac{2\pi}{A}\right) = 10^{0.30}$

$A = 3.1$

b $T = \dfrac{2\pi L^q}{A}$

$5.0 = \dfrac{2\pi L^{0.51}}{3.1}$

$L = \sqrt[0.51]{\dfrac{(5.0 \times 3.1)}{2\pi}}$

$L = 5.9\,\text{m}$

TIP

You could also have found values for L by substituting into $\log T = \log\left(\dfrac{2\pi}{A}\right) + q \log L$ which,

when rearranged, gives $\log L = \dfrac{\log T - \log\left(\dfrac{2\pi}{A}\right)}{q} = \dfrac{\log 5.0 - y \text{ intercept}}{\text{gradient}}$ giving

$L = 10^{\frac{\log 5.0 - y \text{ intercept}}{\text{gradient}}} = 10^{\frac{\log 5.0 - 0.3}{0.51}} = 6.1\,\text{m}$

EXERCISE 7F

a A student investigates how the frequency, f, of a standing wave on a cord varies with the mass, m, suspended from one end of the cord. Their apparatus is shown in Figure 7.3.

The suggested relationship is $f = A m^q$.

A and q are constants. A graph is plotted with $\log f$ on the y-axis and $\log m$ on the x-axis.

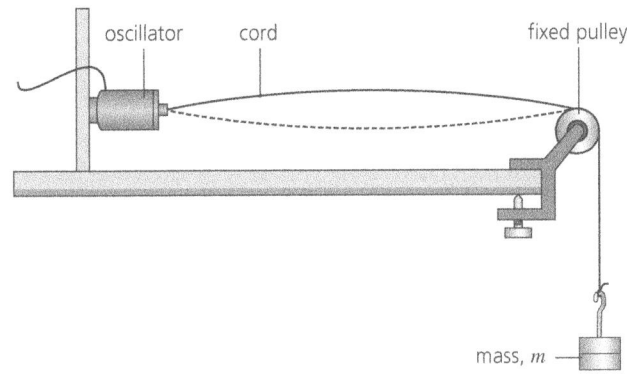

Figure 7.3 Standing wave on a stretched cord

    i The gradient of the graph is 0.50 and the y-intercept is 1.6. Use these values to determine A and q. You do not have to include their absolute uncertainty or their units. [3]

    ii Use your answer to part **i** to determine the frequency, f, when the mass, m, is 2.00 kg. [2]

b A student investigates how the angular amplitude, A, of a pendulum varies with time, t. It is suggested the relationship is $A = pe^{-qt}$ where p and q are constants.

The student plots a graph of $\ln A$ on the y-axis against t on the x-axis. The gradient of their graph is −0.0305 and the y-intercept is 3.22.

    i Use values of the gradient and the y-intercept to determine the values of the constants A and q. [3]

    ii Use your answer to part **i** to determine the amplitude of the oscillation after 20 s. [2]

8 Practice questions

In Paper 5, the questions are based on knowledge and equipment covered in the syllabus.

1 A student is investigating the bending of a loaded wooden strip. Figure 1.1 shows a rectangular strip of width, b, and thickness, t, overhanging the edge of a bench. A length, L, of the strip is unsupported.

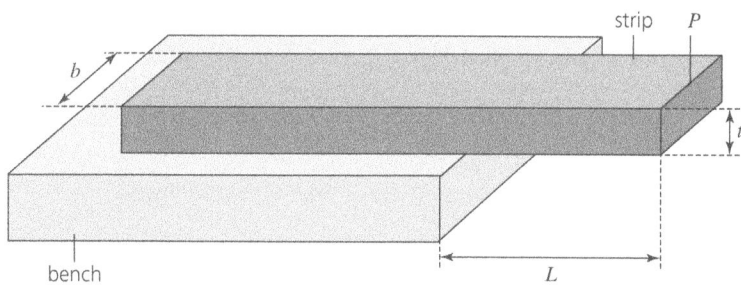

Figure 1.1

A load of mass, M, is positioned at point, P. This causes the unsupported part of the strip to bend with a deflection, s, as shown in Figure 1.2.

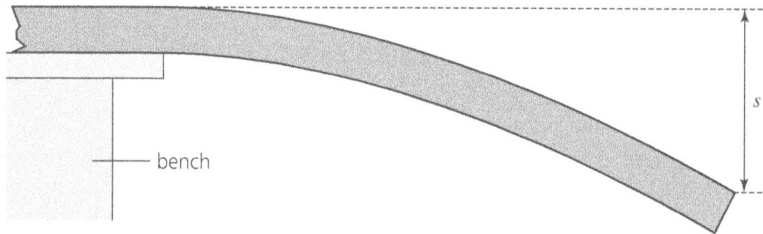

Figure 1.2

It is suggested that the relationship between s and L is

$$E = \frac{4MgL^3}{bst^3}$$

where g is the acceleration of free fall and E is the Young modulus of the wood.

Design a laboratory experiment to test the relationship between s and L.

Explain how your results could be used to determine a value for E.

You should draw a diagram, showing the arrangement of your equipment. In your account you should pay particular attention to:
- the procedure to be followed
- the measurements to be taken
- the control of variables
- the analysis of the data
- any safety precautions to be taken.

[15]

Adapted from Cambridge International AS & A Level Physics 9702 Paper 51 Q1 May/June 2019

Mark this question carefully before trying the next one.

2 A student investigates the motion of a trolley on a wooden surface, as shown in Figure 2.1.

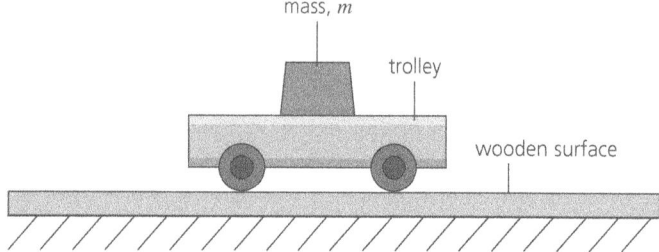

Figure 2.1

A mass, m, is placed on the trolley.

A mass, P, is attached to the trolley by string which passes over a pulley. When this mass falls, it pulls the trolley along the surface.

The trolley is initially at rest. The student investigates how the speed, v, of the trolley at a distance, d, from the initial position of the trolley varies with m.

It is suggested that the relationship between v and m is

$$\frac{2d}{v^2} = \frac{m + R}{Pg - Q}$$

where g is the acceleration of free fall and Q and R are constants.

Design a laboratory experiment to test the relationship between m and v.

Explain how your results could be used to determine values for Q and R.

You should draw a diagram, in the space provided below, showing the arrangement of your equipment. In your account you should pay particular attention to:
- the procedure to be followed
- the measurements to be taken
- the control of variables
- the analysis of the data
- any safety precautions to be taken. [15]

Cambridge International AS & A Level Physics 9702 Paper 52 Q1 October/November 2020

A LEVEL

3 A student investigates stationary sound waves in cylindrical tubes. Figure 3.1 shows a stationary wave pattern in a tube which is open at both ends.

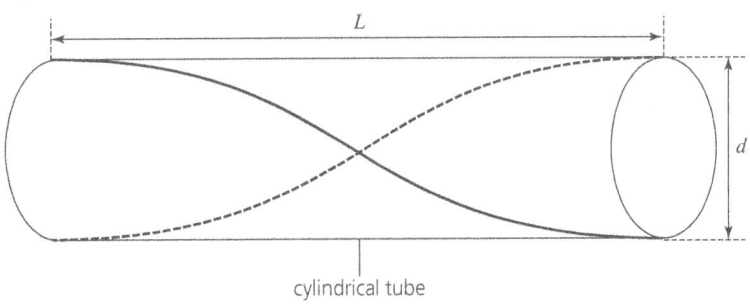

Figure 3.1

The tube has length L and diameter d. The frequency of the sound for the stationary wave pattern shown is f.

There are a number of different tubes available.

It is suggested that the relationship between f and d is

$$\frac{v}{f} = 2L + kd$$

where v is the speed of sound in air and k is a constant.

Design a laboratory experiment to test the relationship between f and d.

Explain how your results could be used to determine values for k and v.

You should draw a diagram, in the space provided below, showing the arrangement of your equipment. In your account you should pay particular attention to:
- the procedure to be followed
- the measurements to be taken
- the control of variables
- the analysis of the data
- any safety precautions to be taken. [15]

Cambridge International AS & A Level Physics 9702 Paper 51 Q1 October/November 2021

A LEVEL

4 A student is investigating how the resistance of a thermistor varies with temperature. The thermistor is placed in water, as shown in Figure 4.1.

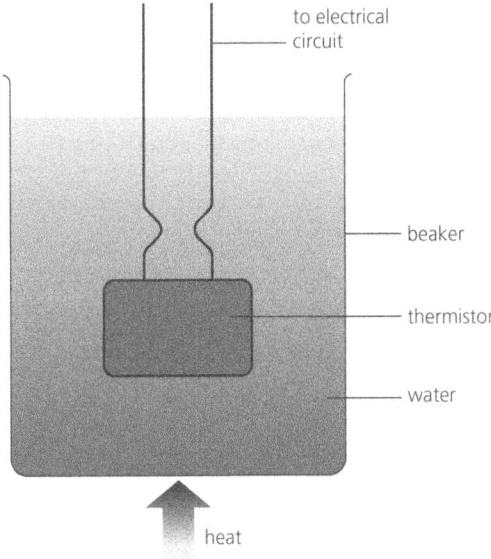

Figure 4.1

The thermistor is connected to a battery with electromotive force (e.m.f.), E, and negligible internal resistance. The current, I, in the thermistor is measured. The resistance, R, of the thermistor is then determined using the expression

$$R = \frac{E}{I}$$

The experiment is repeated for different temperatures of the water.

It is suggested that the resistance, R, of the thermistor and the thermodynamic temperature, T, are related by the equation

$$R = pT^q$$

where p and q are constants.

a A graph is plotted of log R on the y-axis against log T on the x-axis.

Determine expressions for the gradient and the y-intercept.

gradient = ...

y-intercept = ... [1]

b The value of E is 9.4 ± 0.1 V.

Values of T, I and log T are given in Table 4.1

T/K	I/mA	$R/10^3\,\Omega$	log (T/K)	log ($R/10^3\,\Omega$)
303	1.0 ± 0.1		2.481	
313	1.6 ± 0.1		2.496	
323	2.4 ± 0.1		2.509	
333	3.7 ± 0.1		2.522	
343	5.5 ± 0.1		2.535	
353	8.7 ± 0.1		2.548	

Table 4.1

A LEVEL

Calculate and record values of $R/10^3\,\Omega$ and $\log(R/10^3\,\Omega)$ in Table 4.1.

Include the absolute uncertainties in $R/10^3\,\Omega$ and $\log(R/10^3\,\Omega)$. [4]

c i Plot a graph of $\log(R/10^3\,\Omega)$ against $\log(T/K)$.

Include error bars for $\log(R/10^3\,\Omega)$. [2]

ii Draw the straight line of best fit and a worst acceptable straight line on your graph. Both lines should be clearly labelled. [2]

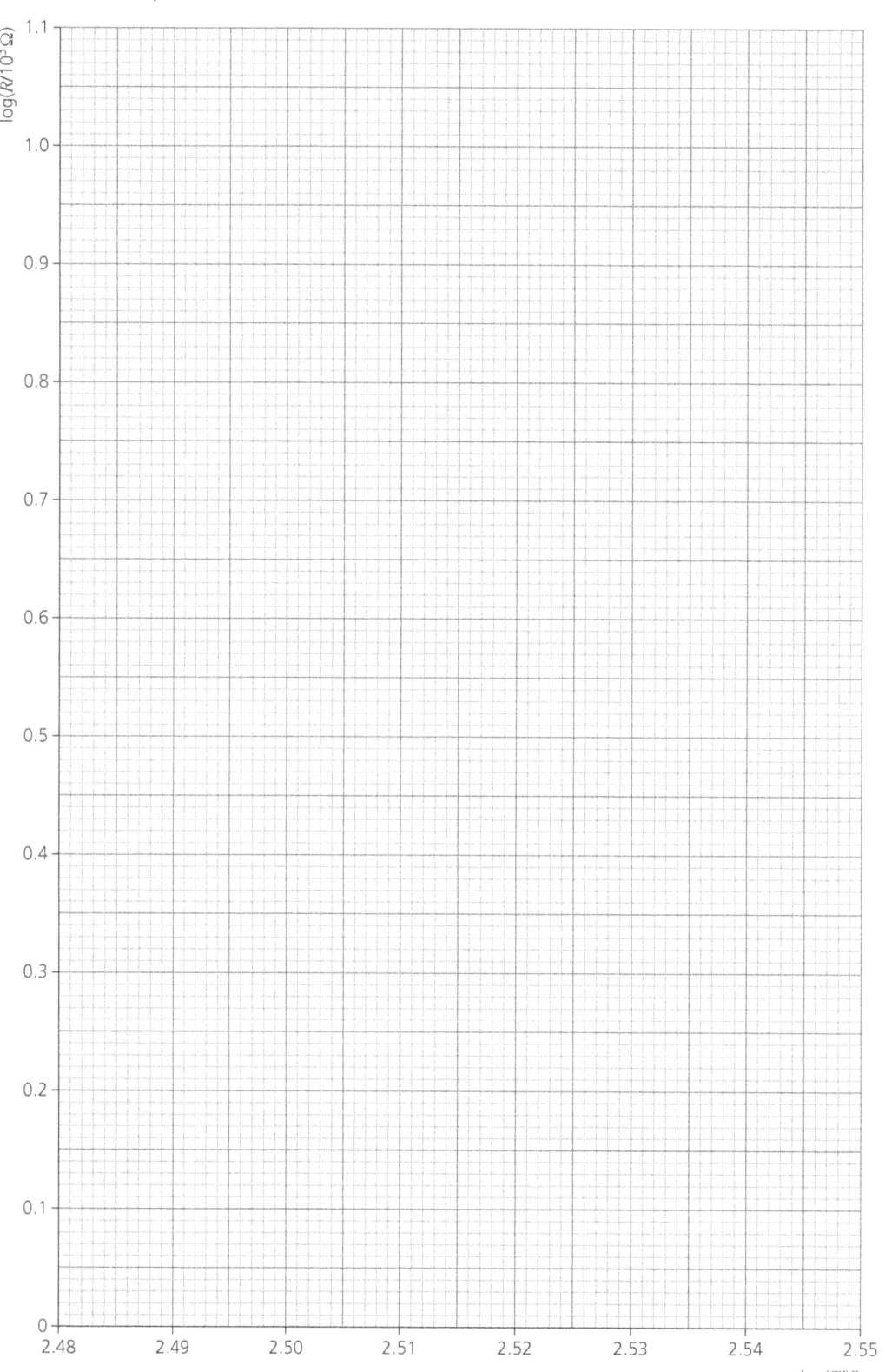

8 Practice questions

iii Determine the gradient of the line of best fit. Include the absolute uncertainty in your answer.

Gradient = ... [2]

iv Determine the *y*-intercept of the line of best fit. Do **not** determine the absolute uncertainty.

y-intercept = ... [1]

d Using your answers to **a**, **ciii** and **civ**, determine the values of *p* and *q*. You need not be concerned with units. Do **not** include the absolute uncertainties.

p = ...
q = ... [2]

e Using your answers to **d**, determine the thermodynamic temperature, *T*, when the resistance of the thermistor is 15 kΩ.

T = ... K [1]

[Total: 15]

Cambridge International AS & A Level Physics 9702 Paper 52 Q2 October/November 2019

A LEVEL

5 A student investigates the discharge of a capacitor through a resistor using the circuit shown in Figure 5.1.

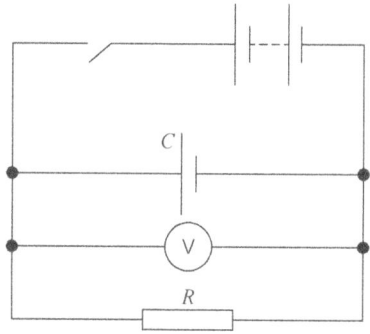

Figure 5.1

The student initially closes the switch and charges the capacitor. The switch is then opened and a stopwatch is started. The capacitor discharges through the resistor. At time, t, the potential difference, V, across the capacitor is measured.

It is suggested that V and t are related by the equation

$$V = \left(\frac{Q_0}{C}\right) e^{-\frac{t}{RC}}$$

where Q_0 is the charge of the fully charged capacitor, C is the capacitance of the capacitor and R is the resistance of the resistor.

a A graph is plotted of $\ln V$ on the y-axis against t on the x-axis.

Determine expressions for the gradient and y-intercept.

gradient = ...

y-intercept = ... [1]

b Values of t and V are given in Table 5.1.

t/s	V/V	$\ln (V/\text{V})$
0	6.2 ± 0.2	
6	4.6 ± 0.2	
12	3.4 ± 0.2	
18	2.6 ± 0.2	
24	2.0 ± 0.2	
30	1.4 ± 0.2	

Table 5.1

Calculate and record values of $\ln (V/\text{V})$ in Table 5.1.

Include the absolute uncertainties in $\ln (V/\text{V})$. [2]

c i Plot a graph of $\ln (V/\text{V})$ against t/s.

Include error bars for $\ln (V/\text{V})$. [2]

ii Draw the straight line of best fit and a worst acceptable straight line on your graph. Both lines should be clearly labelled. [2]

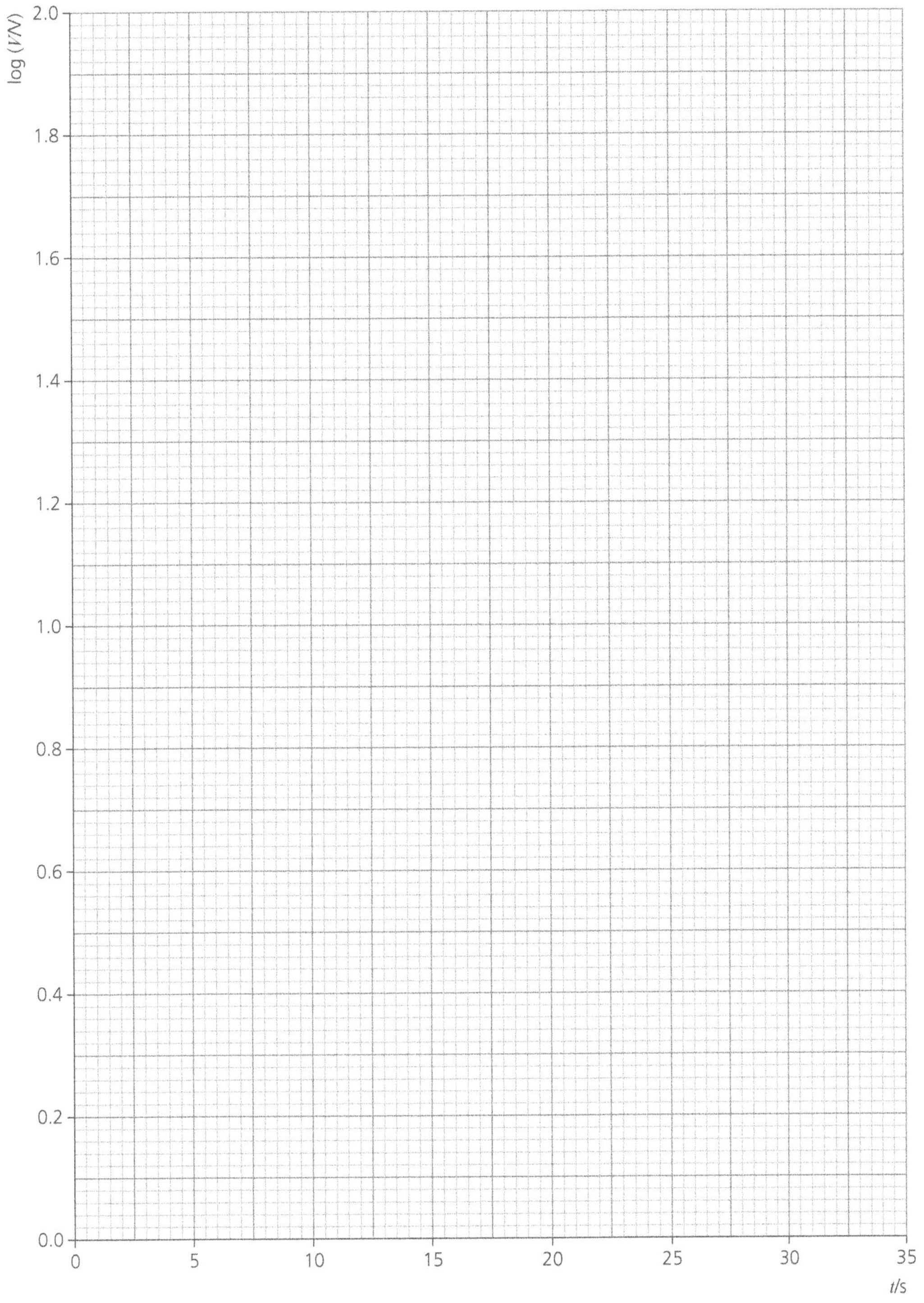

(Graph with y-axis labelled log (V/V) from 0.0 to 2.0 and x-axis labelled t/s from 0 to 35.)

iii Determine the gradient of the line of best fit. Include the absolute uncertainty in your answer.

gradient = ... [2]

iv Determine the y-intercept of the line of best fit. Do not include the absolute uncertainty in your answer.

y-intercept = ... [1]

d i Using your answers to **a**, **ciii** and **civ**, determine values of C and Q_0. Include appropriate units.
Data: $R = 39\,k\Omega$

$C = $ ....

$Q_0 = $ .... [3]

ii The percentage uncertainty in the value of R is 5%.

Determine the absolute uncertainty in C.

absolute uncertainty in $C = $ .... [1]

e Using your results, determine the value of V when the time t is 1.0 minute.

$V = $ .... V [1]

[Total: 15]

Cambridge International AS & A Level Physics 9702 Paper 51 Q2 May/June 2020

6 A student investigates the collision of two gliders A and B on a linear air-track. A card is attached to glider B, as shown in Figure 6.1.

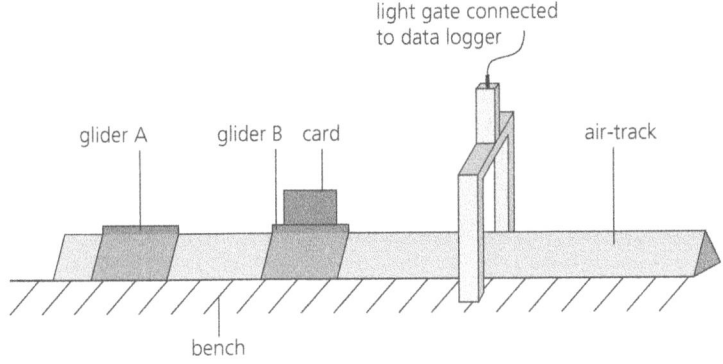

Figure 6.1

Glider B has a mass, M. A mass, m, is added to glider B.

Glider A travels at a constant velocity, u, towards the stationary glider B. The gliders then collide and move together towards the light gate.

The card passes through the light gate which is connected to a data logger. The student records the velocity, v, of the two gliders from the data logger.

The student changes the mass, m, and repeats the experiment.

It is suggested that v and m are related by the equation

$Au = (M + m + A)v$

where A is the mass of glider A.

a A graph is plotted of $\frac{1}{v}$ on the y-axis against $(M + m)$ on the x-axis.

Determine expressions for the gradient and y-intercept.

gradient = ...

y-intercept = ... [1]

b Values of m and v are given in Table 6.1.

The value of M is 330 g $\pm$ 5%.

Each value of m has a percentage uncertainty of $\pm$ 5%.

m/g	$(M+m)$/g	v/cm s^{-1}	$\frac{1}{v}$/s cm^{-1}
50		4.42	
150		3.92	
250		3.40	
350		3.02	
500		2.58	
600		2.33	

Table 6.1

A LEVEL

Calculate and record values of $(M + m)/g$ and $\frac{1}{v}$/s cm^{-1} in Table 6.1.

Include the absolute uncertainties in $(M + m)$. [2]

c i Plot a graph of $\frac{1}{v}$/s cm^{-1} against $(M + m)/g$.

   Include error bars for $(M + m)$. [2]

  ii Draw the straight line of best fit and a worst acceptable straight line on your graph. Both lines should be clearly labelled. [2]

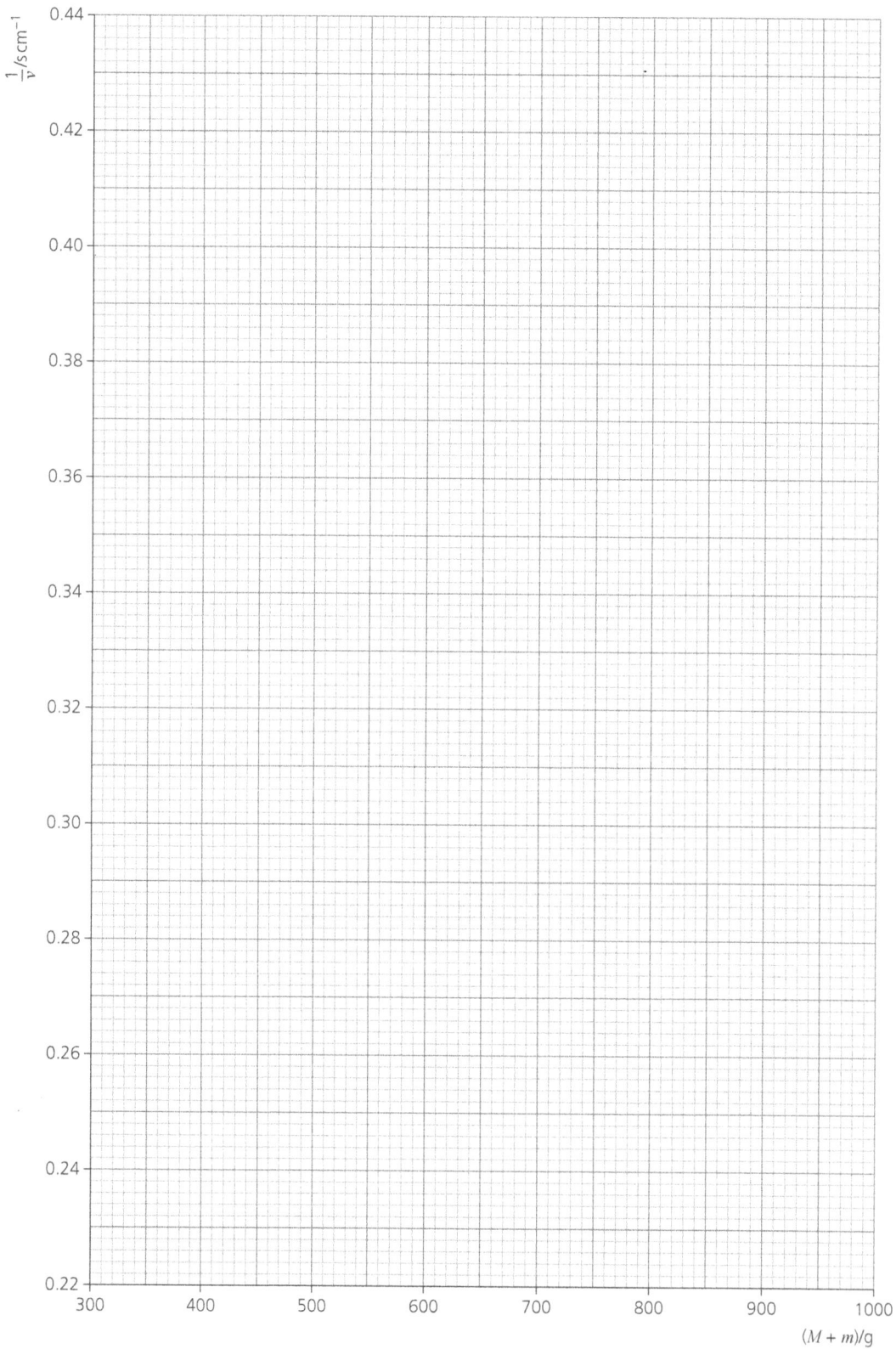

8 Practice questions

iii Determine the gradient of the line of best fit. Include the absolute uncertainty in your answer.

gradient = ... [2]

iv Determine the y-intercept of the line of best fit. Include the absolute uncertainty in your answer.

y-intercept = ... [2]

d i Using your answers to **a**, **ciii** and **civ**, determine the values of u and A. Include appropriate units.

u = ...

A = ... [2]

ii Determine the percentage uncertainty in A.

percentage uncertainty in A = ... % [1]

e The experiment is repeated. Determine the value of m that would give a velocity v of $2.0\,\text{cm s}^{-1}$.

m = ... g [1]

[Total: 15]

Cambridge International AS & A Level Physics 9702 Paper 51 Q2 May/June 2021

Reinforce learning and deepen understanding of the key practical skills required by the Cambridge International AS & A Level Physics (9702) syllabus; an ideal course companion or homework book for use when carrying out and analysing practical work throughout the course.

» Support students' learning and provide guidance on practical skills with extra practice questions and activities, tailored to topics in the Student Book

» Keep track of students' work with ready-to-go write-in exercises which once completed can also be used to recap learning for revision

» Offer extra support for the mathematical and statistical parts of the course

» Answers can be found at www.hoddereducation.com/cambridgeextras

Use with *Cambridge International AS & A Level Physics Student's Book Third Edition*
9781510482807

For over 30 years we have been trusted by Cambridge schools around the world to provide quality support for teaching and learning. For this reason we have been selected by Cambridge Assessment International Education as an official publisher of endorsed material for their syllabuses.

This resource is endorsed by Cambridge Assessment International Education

✓ Provides learner support for the syllabus for examination from 2022

✓ Has passed Cambridge International's rigorous quality-assurance process

✓ Developed by subject experts

✓ For Cambridge schools worldwide

HODDER EDUCATION

e: education@hachette.co.uk
w: hoddereducation.com

ISBN 978-1-5104-8284-5

MIX
Paper | Supporting responsible forestry
FSC™ C104740